Lecture Notes in Computer Scien

Edited by G. Goos, J. Hartmanis, and J. van L

Springer
Berlin
Heidelberg
New York
Barcelona
Hong Kong
London
Milan
Paris
Tokyo

Luca de Alfaro   Stephen Gilmore (Eds.)

Process Algebra and Probabilistic Methods

Performance Modelling and Verification

Joint International Workshop, PAPM-PROBMIV 2001
Aachen, Germany, September 12-14, 2001
Proceedings

 Springer

Series Editors

Gerhard Goos, Karlsruhe University, Germany
Juris Hartmanis, Cornell University, NY, USA
Jan van Leeuwen, Utrecht University, The Netherlands

Volume Editors

Luca de Alfaro
University of California at Berkeley, Dept. of Electr. Eng. and Computer Science
479 Cory Hall, Berkeley, CA 94720-1770, USA
E-mail: dealfaro@eecs.berkeley.edu

Stephen Gilmore
The University of Edinburgh, Laboratory for Foundations of Computer Science
Edinburgh EH9 3JZ, UK
E-mail: stg@dcs.ed.ac.uk

Cataloging-in-Publication Data applied for

Die Deutsche Bibliothek - CIP-Einheitsaufnahme

Process algebra and probabilistic methods : performance modeling and
verification ; joint international workshop ; proceedings / PAPM PROBMIV
2001, Aachen, Germany, September 12 - 14, 2001. Luca de Alfaro ; Stephen
Gilmore (ed.). - Berlin ; Heidelberg ; New York ; Barcelona ; Hong Kong ;
London ; Milan ; Paris ; Tokyo : Springer, 2001
  (Lecture notes in computer science ; Vol. 2165)
  ISBN 3-540-42556-X

CR Subject Classification (1998): F.3.1, F.3, D.2.4, D.3.1, C.4

ISSN 0302-9743
ISBN 3-540-42556-X Springer-Verlag Berlin Heidelberg New York

Springer-Verlag Berlin Heidelberg New York
a member of BertelsmannSpringer Science+Business Media GmbH

http://www.springer.de

© Springer-Verlag Berlin Heidelberg 2001
Printed in Germany

Typesetting: Camera-ready by author, data conversion by PTP-Berlin, Stefan Sossna
Printed on acid-free paper   SPIN: 10840347   06/3142   5 4 3 2 1 0

Preface

This volume contains the proceedings of the first joint PAPM-PROBMIV Workshop, held at the Rheinisch-Westfälische Technische Hochschule (RWTH) Aachen, Germany, 12–14 September 2001.

The PAPM-PROBMIV workshop results from the combination of two workshops: PAPM (Process Algebras and Performance Modeling) and PROBMIV (Probabilistic Methods in Verification). The aim of the joint workshop is to bring together the researchers working across the whole spectrum of techniques for the modeling, specification, analysis, and verification of probabilistic systems. Probability is widely used in the design and analysis of software and hardware systems, as a means to derive efficient algorithms (e.g. randomization), as a model for unreliable or unpredictable behavior (as in the study of fault-tolerant systems and computer networks), and as a tool to study performance and dependability properties. The topics of the workshop include specification, models and semantics of probabilistic systems, analysis and verification techniques, probabilistic methods for the verification of non-probabilistic systems, and tools and case studies.

The first PAPM workshop was held in Edinburgh in 1993; the following ones were held in Regensberg (1994), Edinburgh (1995), Torino (1996), Enschede (1997), Nice (1998), Zaragoza (1999), and Geneva (2000). The first PROBMIV workshop was held in Indianapolis, Indiana (1998); the next one took place in Eindhoven (1999). In 2000, PROBMIV was replaced by a Dagstuhl seminar on Probabilistic Methods in Verification.

The PAPM-PROBMIV workshop is held in conjunction with two other workshops: 11th GI/ITG Conference on Measuring, Modeling, and Evaluation of Computer and Communications Systems (MMB), and the 9th International Workshop on Petri Nets and Performance Models (PNPM). Together, these three workshops form the *2001 Aachen Multiconference on Measurement, Modeling, and Evaluation of Computer-Communication Systems.* We hope that this setting fosters the exchange of ideas with neighboring research fields and allows for a comparison of different viewpoints towards similar problems.

Of the 23 regular papers, 12 were accepted for presentation at the workshop and are included in the present volume. The workshop is preceded by three tutorials, given by Joost-Pieter Katoen (University of Twente) on *Probabilistic verification of Markov chains,* by Marina Ribaudo (University of Torino) on *An introduction to stochastic process algebras,* and by Roberto Segala (University of Bologna) on *Nondeterminism in probabilistic verification.* The workshop includes three invited presentations, by Shankar Sastry (University of California, Berkeley), Markus Siegle (Friedrich-Alexander Universität Erlangen-Nürnberg), and Frits Vaandrager (University of Nijmegen).

We thank all the members of the program committee, and their sub-referees, for selecting the papers to be presented. Special thanks are due to Boudewijn

Haverkort (University of Aachen), the general chair of the multi-conference and local organization, and Peter Kemper (University of Dortmund), the tool session chair. Our thanks go to the following organizations for their generous sponsorship of the Aachen multiconference: German Research Association (DFG), IBM Deutschland, Siemens AG München (Information and Communication Networks), T-Nova Deutsche Telekom Innovationsgesellschaft mbH, and TENOVIS. Our thanks also go to all the authors for meeting the tight deadlines which we set without compromising on the rigor or clarity of their papers.

July 2001   Luca de Alfaro
                                                                Stephen Gilmore

Scientific Organization

PAPM-PROBMIV 2001 Program Committee

Luca de Alfaro (Co-chair, University of California, Berkeley)
Stephen Gilmore (Co-chair, University of Edinburgh)
Gianfranco Ciardo (College of William and Mary)
Rance Cleaveland (SUNY Stony Brook)
Roberto Gorrieri (University of Bologna)
Boudewijn Haverkort (RWTH Aachen)
Ulrich Herzog (University of Erlangen)
Michael Huth (Kansas State University)
Radha Jagadeesan (Loyola University)
Joost-Pieter Katoen (University of Twente)
Leïla Kloul (University of Versailles)
Kim Larsen (University of Aalborg)
Corrado Priami (University of Verona)
Marina Ribaudo (University of Turin)
Roberto Segala (University of Bologna)
Moshe Vardi (Rice University)

General Chair

Boudewijn Haverkort (University of Aachen)

Tool Session Chair

Peter Kemper (University of Dortmund)

PROBMIV Steering Committee

Marta Kwiatkowska (Chair, University of Birmingham)
Luca de Alfaro (University of California, Berkeley)
Rajeev Alur (University of Pennsylvania)
Christel Baier (University of Mannheim)
Michael Huth (Kansas State University)
Joost-Pieter Katoen (University of Twente)
Prakash Panangaden (McGill University)
Roberto Segala (University of Bologna)

PAPM Steering Committee

Ed Brinksma (University of Twente)
Roberto Gorrieri (University of Bologna)
Ulrich Herzog (University of Erlangen)
Jane Hillston (University of Edinburgh)

Referees

Alessandro Aldini	Purush Iyer
Luca de Alfaro	Paul Jackson
Suzana Andova	Radha Jagadeesan
Marco Bernardo	Joost-Pieter Katoen
Henrik Bohnenkamp	Peter Kemper
Jeremy Bradley	Uli Klehmet
Mario Bravetti	Leïla Kloul
Peter Buchholz	Rom Langerak
Gianfranco Ciardo	Kim Larsen
Graham Clark	Joachim Meyer-Kayser
Rance Cleaveland	Gethin Norman
Lucia Cloth	Prakash Panangaden
Pedro d'Argenio	Dave Parker
Jean-Michel Fourneau	Corrado Priami
Stephen Gilmore	Marina Ribaudo
Roberto Gorrieri	Roberto Segala
Marco Gribaudo	Riccardo Sisto
Vineet Gupta	Gene Stark
Boudewijn Haverkort	Nigel Thomas
Holger Hermanns	Joanna Tomasik
Ulrich Herzog	Moshe Vardi
Andras Hovarth	Katinka Wolter
Michael Huth	Wlodek Zuberek

Organization

Sponsors

German Research Association (DFG)
IBM Deutschland
Siemens AG München, Information and Communication Networks
T-Nova Deutsche Telekom Innovationsgesellschaft mbH
TENOVIS

Local Arrangements Committee

Alexander Bell
Henrik Bohnenkamp
Rachid El Abdouni Kayari
Lucia Cloth
Boudewijn Haverkort
Ramin Sadre

Finance Chair

Volker Schanz (ITG, Frankfurt/Main)

Table of Contents

Advances in Model Representations

Markus Siegle

Lehrstuhl für Informatik 7, University of Erlangen-Nürnberg, Germany
siegle@informatik.uni-erlangen.de

Abstract. We review high-level specification formalisms for Markovian performability models, thereby emphasising the role of structuring concepts as realised par excellence by stochastic process algebras. Symbolic representations based on decision diagrams are presented, and it is shown that they quite ideally support compositional model construction and analysis.

1 Introduction

Stochastic models have a long tradition in the areas of performance and dependability evaluation. Since their specification at the level of the Markov chain is tedious and error-prone, several high-level model specification formalisms have been developed, such as queueing networks, stochastic Petri nets and networks of stochastic automata, which allow humans to describe the intended behaviour at a convenient level of abstraction. Although under Markovian assumptions the analysis of the underlying stochastic process does not pose any conceptual problems, the size of the underlying state space often renders models intractable in practice. Structuring concepts have shown to be of great value in order to alleviate this well-known state space explosion problem.

A process algebra is a mathematically founded specification formalism which provides compositional features, such as parallel composition of components, abstraction from internal actions, and the replacing of components by behaviourally equivalent ones. Therefore, stochastic extensions of process algebras are among the methods of choice for constructing complex, hierarchically structured stochastic models.

Recently, decision diagrams, which were originally developed as memory-efficient representations of Boolean functions in the area of hardware verification, have been extended in order to capture the numerical information which is contained in stochastic models. They have already been successfully used as the underlying data structure in prototype tools for performance analysis and verification of probabilistic systems. In this paper, it is shown that symbolic representations based on decision diagrams are particularly attractive if applied in a compositional context, as provided, for example, by a process algebraic specification formalism. In many cases, decision diagrams allow extremely compact representations of huge state spaces, and it has been demonstrated that all steps of model construction, manipulation and analysis (be it model checking, numerical analysis, or a combination of the two) can be carried out on the decision

L. de Alfaro and S. Gilmore (Eds.): PAPM-PROBMIV 2001, LNCS 2165, pp. 1–22, 2001.
© Springer-Verlag Berlin Heidelberg 2001

diagram based representations. Thus, we argue that decision diagrams fit in well with structured modelling formalisms and open new ways towards increasing the range of manageable performance evaluation and verification problems.

This paper does not intend to present new research results, but to survey the history of structured model representations, with special emphasis on process algebras and symbolic encodings. We provide many pointers to further reading, without attempting to be exhaustive.

The paper is organised as follows: In Sec. 2, we survey the evolution from monolithic to modular model specification formalisms. Sec. 3 reviews the concept of stochastic process algebras. Sec. 4 introduces the symbolic representation of Markovian models with the help of multi-terminal binary decision diagrams and describes compositional model construction, manipulation and analysis on the basis of this data structure. The paper concludes with Sec. 5.

2   From Unstructured to Structured Models

2.1   Monolithic Model Representations

Continuous Time Markov Chains (CTMC) are the basic formalism for specifying performance and dependability models[1]. A CTMC consists of a (finite, for our purpose) set of states and a finite set of transitions between states. The transitions are labelled by positive reals, called the transition rates, which determine transition probabilities and state sojourn times (the latter being exponentially distributed). Time-dependent state probabilities can be derived by solving a system of ordinary differential equations, and steady-state probabilities are calculated by solving a linear system of equations (see, for instance [99]). In order to save memory space, CTMCs are commonly represented as sparse matrices, where essentially only the non-zero entries are stored.

The direct specification of a CTMC at the level of individual states and state-to-state transitions is tedious and error-prone, and therefore only feasible for very small models. This motivated researchers to develop high-level specification formalisms for defining Markovian models at a level of abstraction which is more convenient for the human modeller. The most popular of these formalisms are queueing networks and stochastic Petri nets.

Queueing networks (QN), developed mainly in the 1960ies and 1970ies for modelling time-sharing and polling systems, describe customers moving between stations where they receive service after possibly waiting for a service unit to become available. The aim of analysis is typically the mean or distribution of the number of customers at a station, the customer throughput at a station, or the waiting time. The success of queueing networks stems mainly from the fact that for the class of product form networks [5] very efficient analysis algorithms, such as Buzen's algorithm [24] or mean-value analysis [86], are known, and that software tools for the specification and analysis of QN models were available at

---

[1] In this paper, we do not consider the line of research on non-Markovian models such as described, for example, in [47].

an early stage [89,100]. Although QN have been extended in various directions, e.g. in order to model the forking and synchronisation of jobs (fork-join QNs, [3,68,80]), the formalism of QNs is not suitable for the modelling of arbitrary systems, but specialised to the application area of shared resource systems.

Stochastic Petri nets (SPN) were developed in the 1980ies for modelling complex synchronisation schemes which cannot easily be expressed by queueing models [79]. The modelling primitives of Petri nets (places, transitions, markings) are very basic and do not carry any application-specific semantics. For that reason, Petri nets are universally applicable and very flexible, which is reflected by the fact that they have been successfully applied to many different areas of application. In the class of generalised SPNs (GSPN) [1,2], transitions are either timed or immediate. Timed transitions are associated with an exponentially distributed firing time, while immediate transitions fire as soon as they are enabled. During the analysis of a GSPN, the reachability graph is generated and the so-called vanishing markings, which are due to the firing of immediate transitions, are eliminated. The result is a CTMC whose analysis yields (steady-state or transient) state probabilities, i.e. the probabilities of the individual net markings, from which high-level measures can be computed.

Some software tools for performance modelling, e.g. USENUM [90], MARCA [98], MOSEL [8] and DNAmaca [70], implement their own specialised model description languages, which can also be considered as high-level specification formalisms for CTMCs.

With the help of the high-level model specification formalisms considered so far it is possible to specify larger CTMCs than at the state-to-state level, but these formalisms do not support the concepts of modularity, hierarchy or composition of submodels. As a result, the models are monolithic and may be difficult to understand and debug. Moreover, state space generation and numerical analysis of very large monolithic CTMCs is often not feasible in practice due to memory and CPU time limitations, which is referred to as the notorious state space explosion problem.

A large state space may become tractable if it is decomposed into smaller parts [95,33]. Instead of analysing one large system, the decomposition approach relies on analysing several small subsystems, analysing an aggregated overall system, and afterwards combining the subsystems' solutions accordingly. In general, this approach works well for nearly completely decomposable (NCD) systems whose state space can be partitioned into disjoint subsets of states, such that there is a lot of interaction between states belonging to the same subset, but little interaction between states belonging to different subsets. For the class of reversible Markov chains, the decomposition/aggregation approach yields exact results [32]. We mention that the approach may also be applied iteratively [34,25]. The major question is, of course, how to best partition a given state space, and in general this information should be derived from a modular high-level model specification. Approximate decomposition-based analysis methods for stochastic process algebra models (see Sec. 3) are discussed in [73], where time scale decomposition is based on the concept of NCD Markov chains [65,72], and response time approximation relies on a structural decomposition for the special class of decision-free processes [74]. Another such approach, based on the exploitation of the structure of a special class of process algebraic models, is described in [7]. Approximate decomposition-based analysis for nearly-independent GSPN structures is considered in [27,30].

2.2   Modular Model Representations

Queueing models, stochastic Petri nets and the tool-specific modelling languages mentioned above do not offer the possibility of composing an overall model from components which can be specified in isolation. Such a composition, however, is a highly desirable feature when modelling complex systems, since it enables human users to focus on manageable parts from which a whole system can be constructed. For instance, modern performance analysis advocates a separation of the load model and the machine model, an idea developed already in [69,62, 63], and similar ideas are also applied in stochastic rendezvous networks [101] and layered queueing networks [87]. As another, specific example, suppose one wished to model a communication system where two partners communicate over some communication medium. The model should reflect this structure, i.e. it should consist of three interacting submodels, one for each partner, and one for the medium. The user should be able to specify these three submodels more or less independently of each other and then simply specify the way in which they interact.

In the basic GSPN formalism, a model consists of a single net which covers the whole system to be studied. Therefore, GSPN models of complex systems tend to become very large and confused and suffer from the state space explosion problem. Stochastic activity networks [88,35] constitute an approach to the structuring of GSPNs through the sharing of places between different subnets. In the presence of symmetric submodels, they tackle the state space explosion problem by directly generating a reduced reachability graph in which all mutually symmetric markings are collapsed into one. Symmetries also play a predominant role for the analysis of stochastic well-formed coloured Petri nets [26,43], where a reduced reachability graph is constructed directly from the net description, without the need to construct the full reachability graph first. Another line of research is concerned with building SPNs in a structured way, basically by synchronising subnets via common transitions, which is an instance of the Kronecker approach described below.

Stochastic automata networks (SAN)[2], developed in the 1980ies and 1990ies [82,83,84,85], consist of several stochastic automata, basically CTMCs whose transitions are labelled with event names, which run in parallel and may perform certain synchronising events together. Thus, the SAN formalism is truly structured, since it allows the user to specify an overall model as a collection of interacting submodels. The major attraction of SANs is their memory-efficient

---

[2] The acronym SAN is also used for stochastic activity networks (see above), but in this paper it stands for stochastic automata networks.

representation of the generator matrix of the Markov chain underlying the overall model. This so-called Kronecker (or tensor) approach has since been adapted to queueing networks [17], stochastic Petri nets [14,15,16,21,28,39,40], stochastic process algebras [18] and other structured modelling frameworks [19,22,91,92].

The Kronecker approach realises an implicit, space-efficient representation of the transition rate matrix of a stuctured Markov model. Suppose we have two independent CTMCs C_1 and C_2 which are given by their transition rate matrices R_1 and R_2 (of size d_1 and d_2). Let us consider the combined stochastic process C whose state space is the Cartesian product of the state spaces of C_1 and C_2. Process C possesses the transition rate matrix R which is given by the Kronecker sum of R_1 and R_2:

$$R = R_1 \oplus R_2 = R_1 \otimes I_{d_2} + I_{d_1} \otimes R_2$$

where $\otimes$ denotes Kronecker product, $\oplus$ denotes Kronecker sum and I_d denotes an identity matrix of size d [37]. If, however, C_1 and C_2 are not independent, but perform certain transitions synchronously, the expression for the overall transition rate matrix changes to

$$R = R_{1,i} \oplus R_{2,i} + \sum_{a \in S} \lambda_a \cdot R_{1,a} \otimes R_{2,a}$$

where $R_{1,i}$ and $R_{2,i}$ contain those transitions which C_1 and C_2 perform independently of each other, and $R_{1,a}$ and $R_{2,a}$ contain those transitions which are caused by an event a from the set of synchronising events S. Here it is assumed that the resulting rate of the synchronising event a is given by λ_a, i.e. it is a predetermined rate, and matrices $R_{1,a}$ and $R_{2,a}$ are indicator matrices which contain only zeroes and ones. (It is also possible that $R_{1,a}$ and $R_{2,a}$ contain rates, in which case in the above subexpression $\lambda_a \cdot R_{1,a} \otimes R_{2,a}$ has to be replaced by $R_{1,a} \otimes R_{2,a}$. This would mean that the resulting rate of a synchronising event is equal to the product of the rates of the participating processes.) For the general case, where the overall model consists of K submodels, the expression for the overall transition rate matrix is given by

$$R = \bigoplus_{k=1}^{K} R_{k,i} + \sum_{a \in S} \lambda_a \cdot \bigotimes_{k=1}^{K} R_{k,a}$$

The strength of the Kronecker approach lies in its memory-efficiency (it suffices to store a set of matrices of the size of the submodels) and in the fact that for performing numerical analysis, the potentially very large overall transition rate matrix never needs to be constructed or stored explicitly. The compactness of the representation of the transition rate matrix carries over to the generator matrix and to the iteration matrices for some of the common stationary iterative methods. Thus, iterative numerical schemes which rely on matrix-vector multiplication as their basic operation, can be performed directly on the tensor descriptor of the iteration matrix (Plateau [82] used the power method, and Buchholz [13] describes Kronecker-based power, Jacobi, modified Gauss-Seidel,

JOR and modified SOR methods). Efficient algorithms for the multiplication of a vector with a Kronecker descriptor are analysed in [42,97] and in [20], where, however, the authors state that "... all Kronecker-based algorithms are less computationally efficient than a conventional multiplication where [the matrix] R ist stored in sparse format ..." and "This suggests that, in practice, the real advantage of Kronecker-based methods lies exclusively in their large memory savings".

When working with the Kronecker approach, the set of states reachable from the initial state may be only a small subset of the Cartesian product of the involved submodel state spaces. This is known as the "potential versus actual state space" problem. If the actual state space is not known before numerical analysis starts, a probability vector of the size of the potential state space must be allocated, which can waste a considerable amount of memory space and even make the whole analysis impracticable. For that reason, Kronecker-based reachability techniques have been developed, which allow one to work on the actual state space or a limited superset thereof [20,29,67,78].

3   Stochastic Process Algebras

In this section, we briefly review the concept of stochastic process algebras (SPA). Since process algebras feature composition operators that allow one to construct complex specifications from smaller ones, we argue that they quite ideally support the specification and analysis of structured models. Next we define a simple SPA language which supports both Markovian and immediate transitions.

Definition 1. Stochastic process algebra language $\mathcal{L}$
Let Act be the set of valid action names and Pro the set of process names. Let action $\tau \in Act$ denote the internal, invisible action. For $P, P_i \in \mathcal{L}$, $a \in Act$, $S \subseteq Act \setminus \{\tau\}$, and $X \in Pro$, the set $\mathcal{L}$ of valid expressions is definded by the following language elements:

stop	inaction		
$a; P$	immediate prefix	$(a, \lambda); P$	Markovian prefix
$P_1 + P_2$	choice	$P_1\|[S]\|P_2$	parallel composition
hide a **in** P	hiding	X	process instantiation

A set of definitions of the form $X := P$ constitutes a process environment.   ■

With the help of a structured operational semantics, a transition system whose states correspond to process terms can be derived as the semantic model of a process algebraic specification. For a discussion of the full set of semantic rules for Markovian process algebras similar to our language $\mathcal{L}$ we refer the interested reader to the literature, see e.g. [6,48,53,55,64]. Here we only mention two selected rules. The first is the rule for synchronisation of two processes via Markovian transitions which can be written as follows:

$$\frac{P \xrightarrow{b,\lambda} P' \quad Q \xrightarrow{b,\mu} Q'}{P\|[S]\|Q \xrightarrow{b,\phi(\lambda,\mu)} P'\|[S]\|Q'} \quad b \in S$$

Note that this rule is parametric in a function ϕ determining the rate of synchronisation, since different synchronisation policies (minimum, maximum, product, ...) are possible. In the process algebra TIPP [60], ϕ is instantiated by multiplication, since strong bisimilarity (see below) is a congruence with respect to parallel composition and abstraction, provided that ϕ is distributive over summation of real values, see [60,52,55]. Note that the apparent rate construction of PEPA [64] requires a function $\phi(P, Q, \lambda, \mu)$ instead of $\phi(\lambda, \mu)$.

The second rule is the one for hiding in the case of immediate transitions, which states that an immediate transition labelled by a is turned into an internal immediate transition labelled by τ:

$$\frac{P \overset{a}{\dashrightarrow} P'}{\textbf{hide } a \textbf{ in } P \overset{\tau}{\dashrightarrow} \textbf{hide } a \textbf{ in } P'}$$

As we shall see, internal immediate transitions, as generated by this rule, play a key role during the transformation from a transition system to a CTMC. For our stochastic process algebra language $\mathcal{L}$, the resulting semantic model is an extended stochastic labelled transition system (ESLTS):

Definition 2. Extended Stochastic Labelled Transition System (ESLTS)
Let S be a finite set of states. Let $s_0 \in S$ be the initial state. Let Act be a finite set of action labels. Let $\dashrightarrow$ be defined as follows:

$$\dashrightarrow \; \subseteq \; S \times Act \times S$$

Let $\longrightarrow$ be defined as follows:

$$\longrightarrow \; \subseteq \; S \times Act \times \mathbb{R}^{>0} \times S$$

We call $\mathcal{T} = (S, Act, \dashrightarrow, \longrightarrow, s_0)$ an Extended Stochastic Labelled Transition System. If $(x, a, y) \in \dashrightarrow$, we say that there is an immediate a-transition from state x to state y and write $x \overset{a}{\dashrightarrow} y$. If $(x, b, \lambda, y) \in \longrightarrow$, we say that there is a Markovian b-transition from state x to state y with rate λ and write $x \overset{b,\lambda}{\longrightarrow} y$. ∎

Note that in view of the symbolic representation described below, we restricted ourselves to finite-state transition systems. An ESLTS whose set of immediate transitions is empty is called SLTS. An example ESLTS is depicted in Fig. 3 (left). Basically, immediate transitions lead to the existence of vanishing (instable) states. These are states which are left as soon as they are entered, i.e. their sojourn time is zero. Conversely, tangible (stable) states are states whose sojourn time has an exponential distribution, i.e. is strictly positive. For the performability analysis of an SPA model, a CTMC is constructed from the ESLTS and analysed with conventional numerical methods. The CTMC is obtained by hiding of all action labels, elimination of the vanishing states and proper cumulation of all Markovian transitions between a given ordered pair of states.

For a compositional framework, as in the context of stochastic process algebras, we propose to refine the well-known notion of vanishing states in the following way:

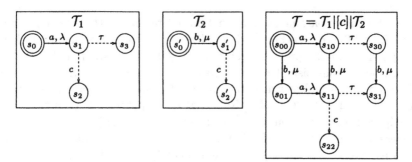

Fig. 1. Role of visible immediate transitions during parallel composition

Definition 3. Compositionally vanishing states
A state s of an ESLTS is called vanishing *if there is at least one internal immediate transition emanating from s (written $s \dashrightarrow^{\tau} s'$). A state s of an ESLTS is called* compositionally vanishing *if it is vanishing and if there is no visible immediate transition emanating from s (written $s \not\dashrightarrow^{a} s''$, where $a \neq \tau$).* ∎

The idea is that even an immediate transition may be delayed if it is visible, since it may be kept waiting by a synchronisation partner which is not yet ready to participate in the synchronisation. Since synchronisation on internal τ-transitions is not allowed, one can be sure that internal immediate transitions will not be delayed. Compositionally vanishing states can be eliminated either before or after composition of subprocesses, but a vanishing state that may also be left by at least one visible immediate transition must not be eliminated before composition[3]. An example for such a situation is shown in Fig. 1 which shows two ESLTSs, $\mathcal{T}_1$ and $\mathcal{T}_2$, which are composed in parallel, synchronising on action c. The resulting ESLTS, $\mathcal{T}$, is shown on the right hand side of the figure. State s_1 in ESLTS $\mathcal{T}_1$, which is vanishing but not compositionally vanishing, must not be eliminated before parallel composition takes place, since its elimination would disable any c-transition in the combined transition system $\mathcal{T}$. In the resulting ESLTS, state s_{10} is a compositionally vanishing state which can be eliminated, whereas state s_{11} is not. However, if action c is hidden in ESLTS $\mathcal{T}$ (since further synchronisation on c is not required), state s_{11} becomes compositionally vanishing and can be also eliminated (its elimination, however, requires a proper treatment of non-determinism as explained below). Note also that there may be one or several Markovian transitions emanating from a vanishing state, but they are never taken. As an example, in Fig. 1 the transition $s_{10} \xrightarrow{b,\mu} s_{11}$ will never be taken, since the competing internal immediate transition $s_{10} \dashrightarrow^{\tau} s_{30}$ will always take place first. Therefore transition $s_{10} \xrightarrow{b,\mu} s_{11}$ can safely be deleted without changing the behaviour of the ESLTS.

---

[3] To complete the picture: A state s is called tangible if there is no immediate transition (i.e. neither visible nor internal) emanating from s. In the remaining case (where there is at least one visible immediate transition, but no internal immediate transition emanating from s) the state is called inconclusive.

The basic strategy of elimination of compositionally vanishing states is to redirect transitions leading to such a state to its successor states. In the case where a compositionally vanishing state has more than one outgoing internal immediate transitions, it is not specified which of them will be taken. This is an instance of non-determinism. In order to resolve such non-determinism, one may assign probabilities or weights to internal immediate transitions. Transitions leading to the compositionally vanishing state can then be redirected to its successor states, taking into account these probabilities.

The concept of bisimilarity is of great importance for SPAs, since it establishes the equivalence between processes, and since it is the basis for state space reduction. Unfortunately, it is beyond the scope of the present paper to discuss bisimulation relations in detail, so we refer to the literature, e.g. [52,57,60,64].

4   Symbolic Representations

In this section we present space-efficient symbolic representations of transition systems with the help of binary decision diagrams (BDD). We review multi-terminal BDDs (MTBDD), also called algebraic decision diagrams, since they are capable of representing real-valued functions [4,31,46].

4.1   Multi-terminal BDDs

Let $\mathbb{B} = \{0,1\}$ denote the set of Booleans[4]. An MTBDD is a graph-based representation of a function $f : \mathbb{B}^n \mapsto \mathbb{R}$.

Definition 4. Multi-Terminal Binary Decision Diagram (MTBDD)
Let $Vars = \{v_1, \ldots, v_n\}$ be a set of Boolean variables with a fixed total ordering $\prec \subset Vars \times Vars$. An (ordered) Multi-Terminal Binary Decision Diagram over $\langle Vars, \prec \rangle$ is a rooted directed acyclic graph $\mathsf{M} = (Vert, \mathsf{var}, \mathsf{else}, \mathsf{then}, \mathsf{value})$ defined by
- *a finite nonempty set of vertices $Vert = T \cup NT$, where T (NT) is the set of terminal (non-terminal) vertices,*
- *a function $\mathsf{var} : NT \mapsto Vars$,*
- *two edge-defining functions $\mathsf{else} : NT \mapsto Vert$ and $\mathsf{then} : NT \mapsto Vert$,*
- *a function $\mathsf{value} : T \mapsto \mathbb{R}$,*
with the following constraints:
$$\forall x \in NT : \mathsf{else}(x) \in T \vee \mathsf{var}(\mathsf{else}(x)) \succ \mathsf{var}(x)$$
$$\forall x \in NT : \mathsf{then}(x) \in T \vee \mathsf{var}(\mathsf{then}(x)) \succ \mathsf{var}(x)$$ ∎

Note that, according to Def. 4, a binary decision tree is an MTBDD. However, we are mainly interested in reduced MTBDDs, defined as follows:

Definition 5. Reducedness of an MTBDD
An MTBDD M is called reduced if and only if the following conditions hold:

---

[4] We use the real numbers 0 and 1 to represent Boolean values, since in the context of MTBDDs Boolean variables will be involved in arithmetic calculations.

1. $\forall x \in NT$: $\mathsf{else}(x) \neq \mathsf{then}(x)$
2. $\forall x, y \in NT$: $x \neq y \Rightarrow (\mathsf{var}(x) \neq \mathsf{var}(y) \vee \mathsf{else}(x) \neq \mathsf{else}(y) \vee \mathsf{then}(x) \neq \mathsf{then}(y))$
3. $\forall x, y \in T$: $x \neq y \Rightarrow \mathsf{value}(x) \neq \mathsf{value}(y)$ ■

The first condition states that there are no "don't care" vertices, i.e. vertices with identical then- and else-successors. The second condition states that there are no two isomorphic non-terminal vertices, and the third condition states that there are no two isomorphic terminal vertices. Bryant [12] proposed a recursive procedure to reduce BDDs[5] that can be applied to MTBDDs as well, and from now on, unless otherwise stated, we assume that MTBDDs are reduced. Fig. 2 (right) shows a reduced MTBDD. In the graphical representation, the edge from a vertex x to $\mathsf{then}(x)$ is drawn solid, and the edge from x to $\mathsf{else}(x)$ is drawn dashed. All vertices that are drawn on one level are labelled with the same Boolean variable, as indicated at the left of the decision diagram. In order to keep the figure clear, all edges leading to the zero-valued terminal vertex are not drawn, i.e. every non-terminal vertex with only one outgoing edge drawn has its other outgoing edge leading to the zero-valued terminal vertex.

Each MTBDD vertex unambiguously defines a real-valued function, based on the so-called Shannon expansion which states that

$$f(v_1, \ldots, v_n) = (1 - v_1) \cdot f(0, v_2, \ldots, v_n) + v_1 \cdot f(1, v_2, \ldots, v_n)$$

The terms $f(0, v_2, \ldots, v_n)$ and $f(1, v_2, \ldots, v_n)$ are called the cofactors of the function f with respect to the Boolean variable v_1.

Definition 6. *Function f_x represented by an MTBDD vertex*
The real-valued function f_x represented by an MTBDD vertex $x \in Vert$ is recursively defined as follows:
- *if $x \in T$ then $f_x = \mathsf{value}(x)$,*
- *else (if $x \in NT$) $f_x = (1 - \mathsf{var}(x)) \cdot f_{\mathsf{else}(x)} + \mathsf{var}(x) \cdot f_{\mathsf{then}(x)}$* ■

Most times one is interested in the case where x corresponds to the MTBDD root. In that case we write f_M instead of f_x, where x is the root vertex of MTBDD M. The two subgraphs of MTBDD M corresponding to the cofactors of f_M are denoted M^{then} and M^{else}, where M^{then} represents $f_M(1, v_2, \ldots, v_n)$ and M^{else} represents $f_M(0, v_2, \ldots, v_n)$. For a fixed ordering of Boolean variables, reduced MTBDDs form a *canonical* representation of real-valued functions, i.e. if M, M′ are two reduced MTBDDs over the same ordered set of Boolean variables $Vars$ such that $f_M = f_{M'}$, then M and M′ are isomorphic.

It should be noted that, given a Boolean function, the size of the resulting MTBDD is highly dependent on the chosen variable ordering. As a prominent example, consider the function $f_{\mathsf{Id}} = \prod_{k=1}^{n} (s_k \equiv t_k)$, which can be interpreted as an identity matrix of size 2^n. Under the interleaved variable ordering $s_1 \prec t_1 \prec \ldots \prec s_n \prec t_n$ the number of vertices needed to represent this function is $3n + 2$, i.e. logarithmic in the size of the matrix. In contrast, using the straight-forward

---

[5] A BDD is an MTBDD where $\forall x \in T$: $\mathsf{value}(x) \in \{0, 1\}$.

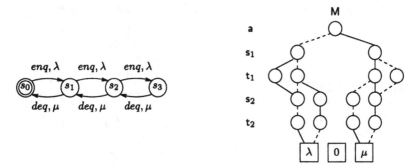

Fig. 2. SLTS and corresponding MTBDD

ordering $s_1 \prec \ldots \prec s_n \prec t_1 \prec \ldots \prec t_n$, the number of vertices is $3 \cdot 2^n - 1$.
Since identity matrices play an important role during the parallel composition
of transition systems (see below), their compact representation is an essential
feature of MTBDDs.

A comprehensive set of logical and arithmetic operations can be realised
efficiently on MTBDDs, such that it is possible to perform calculations on the
functions which are represented by the decision diagrams. The operation APPLY,
for instance, combines two MTBDDs by a binary arithmetic operator, RESTRICT
fixes the value of one or more variables of the MTBDD, and ABSTRACT combines
restricted copies of an MTBDD by an associative binary operator. In general,
algorithms for MTBDD construction and manipulation are variants of their corresponding BDD algorithms [12]. They all follow a recursive descent scheme according to the above Shannon expansion, and their efficiency is achieved through
the clever use of a hash-based vertex table and a cache where intermediate results
are stored for later re-use [11]. MTBDDs are very well suited for the compact
representation of block-structured matrices, and symbolic algorithms for matrix
multiplication and other linear algebra operations exist [4,46,51]. However, existing implementations of MTBDD-based matrix multiplication and vector matrix
multiplication are considerably slower than their sparse counterparts.

4.2 Symbolic Representation of Transition Systems

Fig. 2 shows an SLTS and its symbolic representation by an MTBDD. Since
the set Act of this SLTS only contains two elements, a single Boolean variable
suffices to encode the action label (the case $a = 0$ encodes action enq, and $a = 1$
encodes action deq). Since the SLTS has four states, two bits are required to encode the state identity. We use Boolean variables s_1, s_2 to encoded a transition's
source state, and t_1, t_2 to encode its target state. The transition $s_2 \xrightarrow{enq,\lambda} s_3$, for
example, is encoded by the combination $(a, s_1, t_1, s_2, t_2) = (0, 1, 1, 0, 1)$. Note the
interleaving of the variables for source and target state.

If MTBDD M represents SLTS $\mathcal{T}$ we write $M \rhd \mathcal{T}$. For the symbolic representation of an ESTLS $\mathcal{T}$, one employs two separate decision diagrams, i.e.
an MTBDD M^I which encodes all immediate transitions, and an MTBDD M^M
which encodes all Markovian transitions, as shown in the example of Fig. 3. We

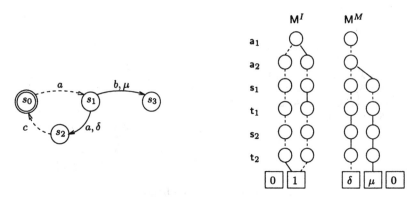

Fig. 3. Encoding of an example ESLTS

then write $(M^I, M^M) \rhd \mathcal{T}$. Basically, M^I is a BDD, since it does not encode any rate values. However, in certain situations one may wish to associate immediate transitions with numerical values, for instance to associate them with weights or probabilities, a feature which may be needed when resolving non-determinism between several concurrently enabled internal immediate transitions. In this case, M^I is a proper MTBDD with possibly more than two terminal vertices. Our tool IM-CAT[6] realises this scheme. A second alternative for the symbolic representation of ESLTSs, where both Markovian and immediate transitions are represented by a single MTBDD, is described in [58].

4.3 Symbolic Manipulation and Analysis of Transition Systems

In this section, we discuss the construction, manipulation and analysis of transition systems represented by MTBDDs. Given a transition system, its symbolic representation can be easily constructed as the sum of the MTBDDs encoding the individual transitions. However, we recommend this procedure only for small transition systems, since the encoding of monolithical transition systems does usually not yield compact representations. Large transition systems should be generated in a compositional fashion from components, following the parallel composition operator of process algebras.

Parallel composition: We now describe MTBDD-based parallel composition, but for simplicity we restrict ourselves to the case of SLTS. The general ESLTS case works in a similar way. Consider the parallel composition of two SLTSs $\mathcal{T}_1$ and $\mathcal{T}_2$ where actions from the set $S \subseteq Act \setminus \{\tau\}$ shall take place in a synchronised way. Using process algebraic notation, we can express this as $\mathcal{T} = \mathcal{T}_1 \|[S]\| \mathcal{T}_2$, where $\mathcal{T}$ is the resulting SLTS. Assume that the MTBDDs which correspond to SLTSs $\mathcal{T}_1$ and $\mathcal{T}_2$ have already been generated and are denoted M_1 and M_2, i.e. $M_i \rhd \mathcal{T}_i$ for $i \in \{1, 2\}$. The set of synchronising actions S can also be encoded in the standard way as a BDD, say S (action labels are encoded by the same Boolean variables in M_1, M_2 and S). The MTBDD M representing

---

[6] IM-CAT is a tool for the compositional construction and analysis of ESLTSs [44,45] which uses the CUDD library [96].

$\mathcal{T}$ (i.e. $\mathsf{M} \triangleright \mathcal{T}$) is constructed as follows:

$$\begin{aligned} \mathsf{M} = \quad & (\mathsf{M}_1 \cdot \mathsf{S}) \cdot (\mathsf{M}_2 \cdot \mathsf{S}) \\ & + \ \mathsf{M}_1 \cdot (1-\mathsf{S}) \cdot \mathsf{Id}_2 \ + \ \mathsf{M}_2 \cdot (1-\mathsf{S}) \cdot \mathsf{Id}_1 \end{aligned}$$

The term on the first line is for the synchronising actions in which both $\mathcal{T}_1$ and $\mathcal{T}_2$ participate. The multiplication $\mathsf{M}_1 \cdot \mathsf{S}$ selects that part of SLTS $\mathcal{T}_1$ which corresponds to actions from the set S, and similarly for $\mathsf{M}_2 \cdot \mathsf{S}$. By then taking the product of these two terms one obtains the encoding of those transitions where both partners simultaneously make a move. The two symmetric terms on the second line are for those actions which $\mathcal{T}_1$ ($\mathcal{T}_2$) performs independently of $\mathcal{T}_2$ ($\mathcal{T}_1$) — these actions are all from the complement of S, encoded by $(1-\mathsf{S})$ — and the multiplication with Id_2 (Id_1) ensures that $\mathcal{T}_2$ ($\mathcal{T}_1$) remains stable, i.e. does not change its state. Note that for the synchronising transitions, calculated by the first line in the above expression, the resulting rate is given by the product $\lambda \cdot \mu$. Should one wish to employ a different function $\phi(\lambda, \mu)$, for instance the maximum function, one would simply have to replace the first line of the above expression by $\text{MAX}(\mathsf{M}_1 \cdot \mathsf{S}, \mathsf{M}_2 \cdot \mathsf{S})$, where MAX is the maximum function on MTBDDs which can be realised with the help of a particular instance of the standard APPLY algorithm.

Enders et al. [41], who considered the parallel composition of BDDs generated from CCS terms, showed that the size of the symbolic representation is proportional to the sum of the sizes of its components, provided that the components are loosely coupled and provided that the interleaved variable ordering is used. We now state a similar result for the parallel composition of ESLTSs. Let $\mathcal{T}_1$ and $\mathcal{T}_2$ be two ESLTSs represented by MTBDDs, i.e. $(\mathsf{M}_i^I, \mathsf{M}_i^M) \triangleright \mathcal{T}_i$ ($i = 1, 2$), using the interleaved variable ordering. Let $(\mathsf{M}^I, \mathsf{M}^M) \triangleright \mathcal{T}_1 \|[S]\| \mathcal{T}_2$ where M^M is constructed from M_1^M, S and M_2^M as above, and M^I is constructed from M_1^I, S and M_2^I in a similar way. Then the number of vertices of M^M is *linear* in the number of vertices of M_1^M and M_2^M, and the number of vertices of M^I is *linear* in the number of vertices of M_1^I and M_2^I. For a proof of this important property see [94].

The fact that parallel composition of components can be realised symbolically in such a way that the size of the data structure grows only linearly compares favourably to the exponential growth resulting from the usual interleaving of causally independent transitions (as generated, for instance, by the operational semantics of process algebras). This feature may be exploited in order to obtain extremely compact representations of huge transition systems. In fact, one can safely state that symbolic representations are only beneficial if they are used in a compositional context.

Reachability analysis: The MTBDD M resulting from the parallel composition of two partners M_1 and M_2 encodes all transitions which are possible in the product space of the two partner processes (called the potential state space). Given a pair of initial states for SLTSs $\mathcal{T}_1$ and $\mathcal{T}_2$, however, only part of this product space (the actual state space) may be reachable due to synchronisation constraints. Therefore, M potentially includes transitions emanating from unreachable states. In this situation, reachability analysis is an important tool for reducing the size of the underlying SLTS. Reachability analysis can be performed efficiently on the symbolic representation of the resulting transition system $\mathcal{T}$ (as described for the purely functional LTS case in [9]), both for SLTSs and ESLTSs represented by MTBDDs.

Hiding: Hiding of action labels, i.e. replacing a visible action a by the internal action τ, can be performed on the MTBDD-based representation of an ESLTS with the help of the operations RESTRICT and APPLY, by selecting and modifying that part of the MTBDD which encodes a-transitions, and by afterwards recombining it with the remaining part of the MTBDD. While the hiding of Markovian transitions does not enable reductions of the transition system, the hiding of immediate transitions may lead to compositionally vanishing states which can be eliminated. In [94], we describe a symbolic algorithm for the elimination of such states that offers a flexible mechanism for resolving non-determinism between several internal immediate transitions, as realised in our tool IM-CAT.

Bisimulation: Symbolic characterisation of strong and weak bisimulation and symbolic algorithms for computing a factorisation of the state space have been described in the literature [9,23,41]. In [61], symbolic algorithms for computing strong and weak Markovian bisimulation on ESLTSs are described in detail, using decision node BDDs (DNBDD) [93], another extension of BDDs for the representation of real-valued functions, as the underlying data structure. These algorithms follow the well-known strategy of iterative refinement and can readily be implemented with the help of MTBDDs.

Numerical analysis: Numerical analysis can be carried out directly on the symbolic representation of the Markov chain [49,50,51]. Direct methods for calculating steady-state probabilities are generally unsuitable for symbolic implementation, since each step modifies the structure of the coefficient matrix and thus the MTBDD structure, which causes considerable overhead to keep the representation canonical and destroys its compactness [4]. For the analysis of large Markov chains based on their symbolic representation, iterative methods are more suitable. Apart from a general matrix powering algorithm [59] that can be instantiated as the power method or the method of Jacobi[7], the projection methods BiCGStab [45,94] and CGS [36] have been realised on the basis of MTBDDs. Unfortunately, the symbolic implementations of these algorithms are all substantially slower than their sparse-matrix counterparts, a fact which is due to the relatively poor performance of symbolic vector-matrix multiplication, as has been observed also in [4,36,38,44,71].

4.4   Compactness of the Symbolic Representation

As an example (taken from [66] and also used in [56]), we consider a cyclic server polling system consisting of d stations and a server. The MTBDD rep-

---

[7] In principle, the method of Gauss-Seidel can also be realised by vector-matrix multiplication, but this requires the inversion of a triangular matrix which usually leads to inefficient encodings.

Table 1. Statistics for the polling system

d	reach. states	transitions	MTBDD size compositional before reachability	MTBDD size compositional after reachability	MTBDD size monolithic
3	36	84	169	203	351
5	240	800	387	563	1,888
7	1,344	5,824	624	1,087	9,056
10	15,360	89,600	1,163	2,459	69,580
15	737,280	6.144e+6	2,191	6,317	–
20	3.14573e+07	3.40787e+08	3,704	13,135	–

resentation of the overall polling model ($\mathcal{T}_{poll}$) is constructed with the help of MTBDD-based parallel composition from $d + 1$ elementary transition systems[8], one for the server ($\mathcal{T}_{serv}$) and one for each station ($\mathcal{T}_{stat_i}$), according to $\mathcal{T}_{poll} := \mathcal{T}_{serv}\|[S]\|(\mathcal{T}_{stat_1}\|[\emptyset]\|\ldots\|[\emptyset]\|\mathcal{T}_{stat_d})$. The order in which the component MTBDDs are generated turned out to be of great importance for the resulting MTBDD size, since it determines the ordering of the MTBDD variables. Unfortunately, it is not obvious a priori which component ordering yields small MTBDDs.

In Tab. 1, the sizes of the resulting MTBDDs are given for different values of d. The first column of the table contains the number of stations d, the 2nd (3rd) column contains the number of reachable states (the number of transitions), and the remaining columns give the number of vertices of the corresponding MTB-DDs[9]. Tab. 1 shows that even for an extremely large state space, the MTBDD representation can be very compact, if it is constructed in a compositional fashion. The last column of Tab. 1 shows the number of MTBDD vertices which one would obtain if one took the monolithic transition system of the overall model as generated by TIPPTOOL (which does not contain unreachable states), and directly encoded it as an MTBDD. Clearly, this method cannot be recommended: Apart from the fact that the transition system of the overall model may not be available due to its excessive size and generation time (as indicated by the "–" entries), the growth of the MTBDD sizes is prohibitive. As expected, the figures in column 4 grow linearly, whereas the ones in column 6 grow exponentially.

The MTBDDs generated compositionally represent all transitions which are possible within the potential state space. As can be observed from the 5th column of Tab. 1, determining the set of reachable states and "deleting" the transitions which originate in unreachable states considerably increases the size of the MTBDDs. Therefore, in general, although MTBDD-based reachability analysis is very fast, it is recommended to work with MTBDDs which represent the potential rather than the actual state space, and store the reachability predicate in a separate BDD. It may seem quite surprising that restriction to the reachable part of the transition system increases the size of the MTBDD. However, similar

---

[8] The elementary transition systems were generated by the stochastic process algebra tool TIPPTOOL [54] and then encoded as individual MTBDDs.

[9] Since the considered version of the polling model does not contain immediate transitions, a single MTBDD (representing Markovian transitions) is sufficient.

phenomena can be observed when performing symbolic elimination of vanishing states or symbolic bisimulation: The size of the symbolic representation grows although the underlying transition system is reduced, i.e. fewer states and transitions are represented. Our explanation for such counter-intuitive behaviour is that the regularity of the model gets lost through the reduction, which destroys the regularity of the MTBBD and thus its compactness (see [59]).

We now mention some figures concerning the MTBDD-based numerical analysis of the polling system: For the case $d = 7$, and working on the potential state space, the MTBDD representing the power iteration matrix, as generated by IM-CAT, has 806 vertices and takes 0.8 seconds to construct[10]. One iteration of the power scheme takes at the average 0.122 seconds, but it takes a ridiculous 8070 power iterations to converge. The MTBDD representing the Jacobi iteration matrix for the same system is larger, it has 1639 vertices and takes 18.94 second to construct, but one Jacobi iteration takes only 0.101 seconds and convergence is reached after 240 iterations. Unfortunately, these speeds are unacceptably slow, if compared to state-of-the-art sparse matrix implementations such as TIPP-TOOL's solver (based on SparseLib1.3 by K. Kundert, UC Berkeley), where one Gauss-Seidel iteration takes only 0.0013 seconds.

5   Discussion and Conclusion

In this paper, we have reviewed the evolution from monolithic to modular model representations. In particular, we have described space-efficient symbolic representations of compositional Markov models stemming from process algebraic specifications, thereby emphasising the role of symbolic parallel composition.

We briefly mention two other data structures related to decision diagrams and developed for the analysis of Markovian systems: Matrix diagrams [28,76, 77,78], an extension of multi-valued decision diagrams, enable the compact representation of structured GSPN models, and probabilistic decision graphs [10] enable a consise representation of probability vectors and probabilistic transition system.

As we have seen, the main bottleneck of the symbolic modelling procedure is numerical analysis. Therefore, speeding up MTBDD-based vector-matrix multiplication remains a major area of research. A promising approach to this problem that combines the advantages of sparse and MTBDD-based representations will be described in [81]. A totally different direction is taken in [75], where special-purpose hardware for the support of basic MTBDD operations has been developed. Parallelisation and distribution of MTBDD manipulation algorithms are also candidates for improving the speed of MTBDD-based numerical analysis.

In summary, we argue that modular model specifications and symbolic representations are a perfect match, and that this combination should play a leading role in future performability analysis and verification projects.

---

[10] The experimental results were obtained on a SUN 5/10 workstation, equipped with 1GB of main memory and running at 300 MHz.

Acknowledgements. The author wishes to thank Holger Hermanns, Marta Kwiatkowska, Gethin Norman and David Parker for many fruitful discussions. Edgar Frank deserves credit for his implementation work on our tool IM-CAT. Thanks also to Lennard Kerber and Matthias Kuntz for critical comments on an earlier version of the paper.

References

1. M. Ajmone Marsan, G. Balbo, and G. Conte. A Class of Generalized Stochastic Petri Nets for the Performance Evaluation of Multiprocessor Systems. *ACM Transactions on Computer Systems*, 2(2):93–122, May 1984.
2. M. Ajmone Marsan, G. Balbo, G. Conte, S. Donatelli, and G. Franceschinis. *Modelling with generalized stochastic Petri nets*. Wiley, 1995.
3. F. Baccelli, W.A. Massey, and D. Towsley. Acyclic Fork–Join Queuing Networks. *Journal of the ACM*, 36(3):615–642, 1989.
4. R.I. Bahar, E.A. Frohm, C.M. Gaona, G.D. Hachtel, E. Macii, A. Pardo, and F. Somenzi. Algebraic Decision Diagrams and their Applications. *Formal Methods in System Design*, 10(2/3):171–206, April/May 1997.
5. F. Baskett, K.M. Chandy, R.R. Muntz, and F.G. Palacios. Open, Closed and Mixed Networks of Queues with Different Classes of Customers. *Journal of the ACM*, 22(2):248–260, 1975.
6. M. Bernardo and R. Gorrieri. A Tutorial on EMPA: A Theory of Concurrent Processes with Nondeterminism, Priorities, Probabilities and Time. *Theoretical Computer Science*, 202:1–54, 1998.
7. H. Bohnenkamp and B. Haverkort. Semi-Numerical Solution of Stochastic Process Algebra Models. In J.-P. Katoen, editor, *ARTS'99*, pages 228–243. Springer, LNCS 1601, 1999.
8. G. Bolch and S. Greiner. Modeling and Performance Evaluation of Production Lines Using the Modeling Language MOSEL. In *Proc. of the 2nd IEEE/ECLA/IFIP Int. Conf. on Architectures and Design Methods for Balanced Automation Systems*, pages 163–174, June 1996.
9. A. Bouali and R. de Simone. Symbolic Bisimulation Minimisation. In *Computer Aided Verification*, pages 96–108, 1992. LNCS 663.
10. M. Bozga and O. Maler. On the Representation of Probabilities over Structured Domains. In N. Halbwachs and D. Peled, editors, *Int. Conf. on Computer-Aided Verification (CAV'99)*, pages 261–273. Springer, LNCS 1633, July 1999.
11. K.S. Brace, R.L. Rudell, and R.E. Bryant. Efficient Implementation of a BDD Package. In *27th ACM/IEEE Design Automation Conf.*, pages 40–45, 1990.
12. R.E. Bryant. Graph-based Algorithms for Boolean Function Manipulation. *IEEE Transactions on Computers*, C-35(8):677–691, August 1986.
13. P. Buchholz. *Die strukturierte Analyse Markovscher Modelle*. PhD thesis, Universität Dortmund, 1991 (in German).
14. P. Buchholz. A Hierarchical View of GCSPNs and Its Impact on Qualitative and Quantitative Analysis. *Journal of Parallel and Distributed Computing*, 15(3):207–224, July 1992.
15. P. Buchholz. Aggregation and Reduction Techniques for Hierarchical GCSPNs. In *Proc. of PNPM '93*, pages 216–225, Tolouse, October 1993.
16. P. Buchholz. Hierarchies in Colored GSPNs. In M. Ajmone Marsan, editor, *14th Int. Conf. on Application and Theory of Petri Nets*, pages 106–125. Springer, LNCS 691, 1993.

17. P. Buchholz. A class of hierarchical queueing networks and their analysis. *Queueing Systems*, 15:59–80, 1994.

18. P. Buchholz. Markovian Process Algebra: Composition and Equivalence. In U. Herzog and M. Rettelbach, editors, *Proc. of the 2nd Workshop on Process Algebras and Performance Modelling*, pages 11–30. Arbeitsberichte des IMMD No. 27/4, Universität Erlangen-Nürnberg, July 1994.

19. P. Buchholz. A Framework for the Hierarchical Analysis of Discrete Event Dynamic Systems. Habilitation thesis, Universität Dortmund, 1996.

20. P. Buchholz, G. Ciardo, S. Donatelli, and P. Kemper. Complexity of memory-efficient Kronecker operations with applications to the solution of Markov models. *INFORMS Journal of Computing*, 12(3):203–222, Summer 2000.

21. P. Buchholz and P. Kemper. On Generating a Hierarchy for GSPN Analysis. *Performance Evaluation Review (ACM Sigmetrics)*, 26(2):5–14, August 1998.

22. P. Buchholz and P. Kemper. A Toolbox for the Analysis of Discrete Event Dynamic Systems. In N. Halbwachs and D. Peled, editors, *Computer Aided Verification*, pages 483–486. Springer, LNCS 1633, 1999.

23. J.R. Burch, E.M. Clarke, K.L. McMillan, D.L. Dill, and L.J. Hwang. Symbolic Model Checking: 10^{20} States and Beyond. *Information and Computation*, (98):142–170, 1992.

24. J.P. Buzen. Computational Algorithms for Closed Queueing Networks with Exponential Servers. *Communications of the ACM*, 16:527–531, 1973.

25. W.L. Cao and W.J. Stewart. Iterative Aggregation/Disaggregation Techniques for Nearly Uncoupled Markov Chains. *Journal of the ACM*, 32(3):702–719, July 1985.

26. G. Chiola, C. Dutheillet, G. Franceschinis, and S. Haddad. On Well-Formed Coloured Nets and their Symbolic Reachability Graph. In *Proc. of the 11th Int. Conf. on Application and Theory of Petri Nets*, pages 387–410, Paris, June 1990. Reprinted in High-level Petri Nets, K. Jensen, G. Rozenberg, eds., Springer 1991.

27. G. Ciardo. *Analysis of Large Stochastic Petri Net Models*. PhD thesis, Duke University, Durham, NC, USA, 1989.

28. G. Ciardo and A.S. Miner. A data structure for the efficient Kronecker solution of GSPNs. In P. Buchholz and M. Silva, editors, *PNPM'99*, pages 22–31. IEEE computer society, 1999.

29. G. Ciardo and M. Tilgner. Parametric State Space Structuring. Technical Report ICASE Report No. 97-67, ICASE, 1997.

30. G. Ciardo and K.S. Trivedi. A decomposition approach for stochastic reward net models. *Performance Evaluation*, 18(1):37–59, July 1993.

31. E.M. Clarke, K.L. McMillan, X. Zhao, M. Fujita, and J. Yang. Spectral Transforms for Large Boolean Functions with Applications to Technology Mapping. In *30th Design Automation Conf.*, pages 54–60. ACM/IEEE, 1993.

32. A.E. Conway and N.D. Georganas. *Queueing Networks – Exact Computational Algorithms*. MIT Press, 1989.

33. P.J. Courtois. *Decomposability, queueing and computer system applications*. ACM monograph series, 1977.

34. P.J. Courtois. On Time and Space Decomposition of Complex Structures. *Communications of the ACM*, 28(6):590–603, June 1985.

35. J. Couvillion, R. Freire, R. Johnson, W.D. Obal II, M.A. Qureshi, M. Rai, W.H. Sanders, and J. Tvedt. Performability modeling with UltraSAN. *IEEE Software*, 8(5):69–80, September 1991.

36. I. Davies. Symbolic techniques for the performance analysis of generalised stochastic Petri nets. Master's thesis, University of Cape Town, Department of Computer Science, January 2001.
37. M. Davio. Kronecker Products and Shuffle Algebra. *IEEE Transactions on Computers*, C-30(2):116–125, February 1981.
38. L. de Alfaro, M. Kwiatkowska, G. Norman, D. Parker, and R. Segala. Symbolic Model Checking for Probabilistic Processes using MTBDDs and the Kronecker Representation. In S. Graf and M. Schwartzbach, editors, *TACAS'2000*, pages 395–410, Berlin, 2000. Springer, LNCS 1785.
39. S. Donatelli. Superposed stochastic automata: a class of stochastic Petri nets with parallel solution and distributed state space. *Performance Evaluation*, 18(1):21–36, July 1993.
40. S. Donatelli. Superposed Generalized Stochastic Petri Nets: Definition and Efficient Solution. In M. Silva, editor, *15th Int. Conf. on Application and Theory of Petri Nets*, Zaragoza, June 1994.
41. R. Enders, T. Filkorn, and D. Taubner. Generating BDDs for symbolic model checking in CCS. *Distributed Computing*, (6):155–164, 1993.
42. P. Fernandes, B. Plateau, and W.J. Stewart. Efficient Descriptor-Vector Multiplications in Stochastic Automata Networks. *Journal of the ACM*, 45(3):381–414, May 1998.
43. G. Franceschinis and M. Ribaudo. Efficient Performance Analysis Techniques for Stochastic Well-Formed Nets and Stochastic Process Algebras. In W. Reisig and G. Rozenberg, editors, *Lectures on Petri Nets II: Applications*, pages 386–437. Springer, LNCS 1492, 1998.
44. E. Frank. Codierung und numerische Analyse von Transitionssystemen unter Verwendung von MTBDDs. Student's thesis, Universität Erlangen–Nürnberg, October 1999 (in German).
45. E. Frank. Erweiterung eines MTBDD-basierten Werkzeugs für die Analyse stochastischer Transitionssysteme. Technical Report Inf 7, 01/00, Universität Erlangen–Nürnberg, January 2000 (in German).
46. M. Fujita, P. McGeer, and J.C.-Y. Yang. Multi-terminal Binary Decision Diagrams: An efficient data structure for matrix representation. *Formal Methods in System Design*, 10(2/3):149–169, April/May 1997.
47. R. German. *Performance Analysis of Communication Systems — Modelling with Non-Markovian Stochastic Petri Nets*. Wiley, 2000.
48. N. Götz, U. Herzog, and M. Rettelbach. Multiprocessor and Distributed System Design: The Integration of Functional Specification and Performance Analysis Using Stochastic Process Algebras. In *Proc. of PERFORMANCE 1993, Tutorial*, pages 121–146. Springer LNCS 729, 1993.
49. G.D. Hachtel, E. Macii, A. Pardo, and F. Somenzi. Probabilistic Analysis of Large Finite State Machines. In *31st Design Automation Conf.*, pages 270–275, San Diego, CA, June 1994. ACM/IEEE.
50. G.D. Hachtel, E. Macii, A. Pardo, and F. Somenzi. Symbolic Algorithms to Calculate Steady-State Probabilities of a Finite State Machine. In *European Design Automation Conf.*, pages 214–218, Paris, February 1994. IEEE.
51. G.D. Hachtel, E. Macii, A. Pardo, and F. Somenzi. Markovian Analysis of Large Finite State Machines. *IEEE Transactions on CAD*, 15(12):1479–1493, Dec. 1996.
52. H. Hermanns. *Interactive Markov Chains*. PhD thesis, Universität Erlangen–Nürnberg, September 1998. Arbeitsberichte des IMMD No. 32/7.
53. H. Hermanns, U. Herzog, and J.-P. Katoen. Process algebra for performance evaluation. *Theoretical Computer Science*, 2001. to appear.

54. H. Hermanns, U. Herzog, U. Klehmet, V. Mertsiotakis, and M. Siegle. Compositional performance modelling with the TIPPtool. *Performance Evaluation*, 39(1-4):5–35, January 2000.

55. H. Hermanns, U. Herzog, and V. Mertsiotakis. Stochastic Process Algebras - Between LOTOS and Markov Chains. *Computer Networks and ISDN systems (CNIS)*, 30(9-10):901–924, 1998.

56. H. Hermanns, J.-P. Katoen, J. Meyer-Kayser, and M. Siegle. A Markov Chain Model Checker. In S. Graf and M. Schwartzbach, editors, *TACAS'2000*, pages 347–362, Berlin, 2000. Springer, LNCS 1785.

57. H. Hermanns, J.-P. Katoen, J. Meyer-Kayser, and M. Siegle. Towards model checking stochastic process algebra. In W. Grieskamp, T. Santen, and B. Stoddart, editors, *2nd Int. Conf. on Integrated Formal Methods*, pages 420–439, Dagstuhl, November 2000. Springer, LNCS 1945.

58. H. Hermanns, M. Kwiatkowska, G. Norman, D. Parker, and M. Siegle. On the use of MTBDDs for performability analysis and verification of stochastic systems. (in preparation).

59. H. Hermanns, J. Meyer-Kayser, and M. Siegle. Multi Terminal Binary Decision Diagrams to Represent and Analyse Continuous Time Markov Chains. In B. Plateau, W.J. Stewart, and M. Silva, editors, *3rd Int. Workshop on the Numerical Solution of Markov Chains*, pages 188–207. Prensas Universitarias de Zaragoza, 1999.

60. H. Hermanns and M. Rettelbach. Syntax, Semantics, Equivalences, and Axioms for MTIPP. In U. Herzog and M. Rettelbach, editors, *Proc. of the 2nd Workshop on Process Algebras and Performance Modelling*, pages 71–88. Arbeitsberichte des IMMD No. 27/4, Universität Erlangen-Nürnberg, July 1994.

61. H. Hermanns and M. Siegle. Bisimulation Algorithms for Stochastic Process Algebras and their BDD-based Implementation. In J.-P. Katoen, editor, *ARTS'99, 5th Int. AMAST Workshop on Real-Time and Probabilistic Systems*, pages 144–264. Springer, LNCS 1601, 1999.

62. U. Herzog. Leistungsbewertung und Modellbildung für Parallelrechner. *Informationstechnik (it)*, 31(1):31–38, 1989. (in German).

63. U. Herzog. Performance Evaluation and Formal Description. In V.A. Monaco and R. Negrini, editors, *Advanced Computer Technology, Reliable Systems and Applications, Proceedings*, pages 750–756. IEEE Comp. Soc. Press, 1991.

64. J. Hillston. *A Compositional Approach to Performance Modelling*. Cambridge University Press, 1996.

65. J. Hillston and V. Mertsiotakis. A Simple Time Scale Decomposition Technique for Stochastic Process Algebras. *The Computer Journal*, 38(7):566–577, December 1995. Special issue: Proc. of the 3rd Workshop on Process Algebras and Performance Modelling.

66. O.C. Ibe and K.S. Trivedi. Stochastic Petri Net Models of Polling Systems. *IEEE Journal on Selected Areas in Communications*, 8(9):1649–1657, December 1990.

67. P. Kemper. Reachability analysis based on structured representation. In J. Billington and W. Reisig, editors, *Application and Theory of Petri Nets*, pages 269–288. Springer, LNCS 1091, 1999.

68. C. Kim and A.K. Agrawala. Analysis of the Fork–Join Queue. *IEEE Transactions on Computers*, 38(2):250–255, 1989.

69. W. Kleinöder. *Stochastische Bewertung von Aufgabenstrukturen für hierarchische Mehrrechnersysteme*. PhD thesis, Universität Erlangen–Nürnberg, Arbeitsberichte des IMMD No. 15/10, August 1982 (in German).

70. W. Knottenbelt. Generalized Markovian Analysis of Timed Transition Systems. Master's thesis, University of Cape Town, June 1996.

71. M. Kwiatkowska, G. Norman, and R. Segala. Automated Verification of a Randomised Distributed Consensus Protocol Using Cadence SMV and PRISM. Technical Report CSR-01-1, School of Computer Science, University of Birmingham, January 2001.

72. V. Mertsiotakis. Time Scale Decomposition of Stochastic Process Algebra Models. In E. Brinksma and A. Nymeyer, editors, *Proc. of 5th Workshop on Process Algebras and Performance Modelling*. CTIT Technical Report series, No. 97-14, University of Twente, June 1997.

73. V. Mertsiotakis. *Approximate Analysis Methods for Stochastic Process Algebras*. PhD thesis, Universität Erlangen-Nürnberg, 1998.

74. V. Mertsiotakis and M. Silva. Throughput Approximation of Decision Free Processes Using Decomposition. In *Proc. of the 7th Int. Workshop on Petri Nets and Performance Models*, pages 174–182, St. Malo, June 1997. IEEE CS-Press.

75. M. Meyer. Entwurf eines spezialisierten Coprozessors für die Manipulation von Binären Entscheidungsdiagrammen. Student's thesis, Universität Erlangen–Nürnberg, January 2001 (in German).

76. A. Miner. Efficient solution of GSPNs using matrix diagrams. In *Petri Nets and Performance models (PNPM)*. IEEE Computer Society Press, 2001. (to appear).

77. A. Miner and G. Ciardo. Efficient reachability set generation and storage using decision diagrams. In H. Kleijn and S. Donatelli, editors, *Application and Theory of Petri Nets 1999*, pages 6–25. Springer, LNCS 1639, 1999.

78. A.S. Miner, G. Ciardo, and S. Donatelli. Using the exact state space of a Markov model to compute approximate stationary measures. *Performance Evaluation Review*, 28(1):207–216, June 2000. Proc. of ACM SIGMETRICS 2000.

79. M.K. Molloy. Performance Analysis Using Stochastic Petri Nets. *IEEE Transactions on Computers*, C-31:913–917, September 1982.

80. R. Nelson and A.N. Tantawi. Approximate Analysis of Fork/Join Synchronization in Parallel Queues. *IEEE Transactions on Computers*, 37(6):739–743, 1988.

81. D. Parker. *Implementation of symbolic model checking for probabilistic systems*. PhD thesis, School of Computer Science, University of Birmingham, 2001. (to appear).

82. B. Plateau. On the Synchronization Structure of Parallelism and Synchronization Models for Distributed Algorithms. In *Proc. of ACM SIGMETRICS*, pages 147–154, Austin, TX, August 1985.

83. B. Plateau and K. Atif. Stochastic Automata Network for Modeling Parallel Systems. *IEEE Transactions on Software Engineering*, 17(10):1093–1108, 1991.

84. B. Plateau and J.-M. Fourneau. A Methodology for Solving Markov Models of Parallel Systems. *Journal of Parallel and Distributed Computing*, 12:370–387, 1991.

85. B. Plateau, J.-M. Fourneau, and K.-H. Lee. PEPS: A Package for Solving Complex Markov Models of Parallel Systems. In *Proc. of the 4th Int. Conf. on Modelling Techniques and Tools for Computer Performance Evaluation*, pages 341–360, Palma (Mallorca), September 1988.

86. M. Reiser and S. Lavenberg. Mean Value Analysis of Closed Multichain Queueing Networks. *Journal of the ACM*, 27(2):313–322, 1980.

87. J.A. Rolia and K.C. Sevcik. The Method of Layers. *IEEE Transactions on Software Engineering*, 21(8):689–700, August 1995.

88. W.H. Sanders and J.F. Meyer. Reduced Base Model Construction Methods for Stochastic Activity Networks. *IEEE Journal on Selected Areas in Communications*, 9(1):25–36, January 1991.

89. C. Sauer and E. McNair. The Evolution of the Research Queueing Package RESQ. In *Proc. of the First Int. Conf. on Modelling Techniques and Tools for Computer Performance Evaluation*, Paris, May 1984.

90. M. Sczittnick. Techniken zur funktionalen und quantitativen Analyse von Markoffschen Rechensystemmodellen. Master's thesis, Universität Dortmund, August 1987 (in German).

91. M. Siegle. Using Structured Modelling for Efficient Performance Prediction of Parallel Systems. In G.R. Joubert, D. Trystram, F.J. Peters, and D.J. Evans, editors, *Parallel Computing: Trends and Applications, Proc. of the Int. Conf. ParCo93*, pages 453–460. North-Holland, 1994.

92. M. Siegle. *Beschreibung und Analyse von Markovmodellen mit großem Zustandsraum*. PhD thesis, Universität Erlangen–Nürnberg, 1995 (in German).

93. M. Siegle. Compositional Representation and Reduction of Stochasitic Labelled Transition Systems based on Decision Node BDDs. In D. Baum, N. Müller, and R. Rödler, editors, *MMB'99*, pages 173–185, Trier, September 1999. VDE Verlag.

94. M. Siegle. Behaviour analysis of communication systems: Stochastic modelling and analysis. Habilitation thesis, University of Erlangen-Nürnberg, 2001 (to appear).

95. H.A. Simon and A. Ando. Aggregation of Variables in Dynamic Systems. *Econometrica*, 29:111–138, 1961.

96. F. Somenzi. CUDD: Colorado University Decision Diagram Package, Release 2.3.0. User's Manual and Programmer's Manual, September 1998.

97. W. Stewart, K. Atif, and B. Plateau. The Numerical Solution of Stochastic Automata Networks. Rapport Apache 6, Institut IMAG, LGI, LMC, Grenoble, November 1993.

98. W.J. Stewart. MARCA: Markov Chain Analyzer, A Software Package for Markov Modeling. In W.J. Stewart, editor, *Numerical Solution of Markov Chains*. Marcel Dekker, 1991.

99. W.J. Stewart. *Introduction to the numerical solution of Markov chains*. Princeton University Press, 1994.

100. M. Veran and D. Potier. QNAP2: A Portable Environment for Queueing Systems Modelling. In *Proc. of the First Int. Conf. on Modelling Techniques and Tools for Computer Performance Evaluation*, Paris, May 1984.

101. M. Woodside, J.E. Neilson, D.C. Petriu, and S. Majumdar. The Stochastic Rendezvous Network Model for Performance of Synchronous Client-Server-like Distributed Software. *IEEE Transactions on Computers*, 44(1):20–34, January 1995.

Faster and Symbolic CTMC Model Checking*

Joost-Pieter Katoen[1], Marta Kwiatkowska[2],
Gethin Norman[2], and David Parker[2]

[1] Formal Methods and Tools Group, Faculty of Computer Science
University of Twente, P.O. Box 217, 7500 AE Enschede, The Netherlands
[2] School of Computer Science, University of Birmingham
Edgbaston, Birmingham B15 2TT, United Kingdom

Abstract. This paper reports on the implementation and the experiments with symbolic model checking of continuous-time Markov chains using multi-terminal binary decision diagrams (MTBDDs). Properties are expressed in Continuous Stochastic Logic (CSL) [7] which includes the means to express both transient and steady-state performance measures. We show that all CSL operators can be treated using standard operations on MTBDDs, thus allowing a rather straightforward implementation of symbolic CSL model checking on existing MTBDD-based platforms such as the verifier PRISM. The main result of the paper is an improvement of $\mathcal{O}(N)$ in the time complexity of checking time-bounded until-formulas, where N is the number of states in the CTMC under consideration. This result yields a drastic speed-up in the verification time of model checking CTMCs, both in the symbolic and non-symbolic case.

1   Introduction

In model-based performance and dependability evaluation, techniques such as stochastic Petri nets, stochastic process algebras, stochastic activity networks, and queueing networks are used to specify the system behaviour at a high level of abstraction. Most of these techniques assume a continuous-time Markov chain (CTMC) as underlying stochastic process. While the analysis of CTMCs focuses mostly on transient-state and steady-state (i.e. long run) characteristics, the specification and analysis of path measures is a subject of growing interest [25].

The temporal logic CSL (Continuous Stochastic Logic) developed originally by Aziz *et al.* [2,3] and extended by Baier *et al.* [7] provides a powerful means to specify path-based as well as traditional state-based measures on CTMCs in a concise, flexible and unambiguous way. CSL is based on the well-known branching-time temporal logic CTL (Computation Tree Logic [11]) and PCTL (Probabilistic CTL [17]); a steady-state operator, a time-bounded until, and a probabilistic (path) operator constitute its main ingredients. It allows one to state, for example, that the probability of reaching a certain set of goal-states within a specified real-valued time bound, provided that all paths to these states obey certain properties, is at least/at most some probability value.

* Partly supported by EPSRC grants GR/M04617, GR/M13046 and GR/N31573.

L. de Alfaro and S. Gilmore (Eds.): PAPM-PROBMIV 2001, LNCS 2165, pp. 23–38, 2001.

Verification of a given finite-state CTMC against a CSL formula is performed using model checking. The model checking problem for CSL is decidable for rational time bounds [2,3]. Approximate CSL model-checking algorithms have been studied in [7] where the satisfaction of time-bounded until formulas is shown to be based on solving a (recursive) Volterra equation system. More recently, Baier *et al.* [6] reduced verifying time-bounded until formulas to the problem of computing transient-state probabilities for CTMCs. This significant result employs a formula-dependent transformation of the CTMC and – more importantly – allows one to adopt efficient techniques like *uniformisation* [16,23] for verifying time-bounded until-formulas. This paper builds upon this earlier work, and considers two issues: improving the time and the space efficiency of model checking CTMCs against CSL formulas based on transient analysis.

Faster CSL model checking. Verifying time-bounded until formulas using transient analysis of CTMCs [6] suggests that uniformisation should be applied to each individual state separately. This results in a worst case time complexity of $\mathcal{O}(M \cdot N)$, where N is the number of states in the CTMC and M the number of transitions. The main result of this paper is an improvement of $\mathcal{O}(N)$ in the time complexity of checking time-bounded until formulas. Inspired by PCTL model checking, the basic idea underlying this efficiency improvement is to carry out the uniformisation for all states at once. Our experiments show that this result yields a drastic speed-up in the verification time of model checking CTMCs.

Symbolic CSL model checking. To combat the infamous state-space explosion problem we investigate representing the state space by multi-terminal binary decision diagrams (MTBDDs [12], also called algebraic decision diagrams [4]). MTBDDs are variants of BDDs that can efficiently deal with real matrices; they allow arbitrary real numbers in the terminal nodes instead of just 0 and 1. We show that CSL model checking can be treated using standard operations on MTBDDs, thus generalising the result for PCTL [5,18] to the continuous-time setting. This basically follows from the fact that CSL model checking amounts to the analysis of either the *embedded* discrete-time Markov chain (DTMC) – in the case of untimed until formulas and steady-state formulas – or the *uniformised* DTMC – in the case of time-bounded until – of the CTMC under consideration. This reduces to graph analysis and iterative matrix-vector multiplication which can be implemented with standard MTBDD operations. Variants of MTBDDs tailored to numerical integration [7] are not needed. This paper reports on the implementation of symbolic CSL model checking as part of PRISM[1] (PRobabilistIc Symbolic Model checker), a prototype tool for the symbolic verification of Markov decision processes (MDPs) with DTMCs as a subset thereof.

Organisation of the paper. Section 2 briefly recalls PCTL model checking. Section 3 introduces CTMCs and CSL. Handling time-bounded until is covered in Section 4. Section 5 discusses CSL model checking with MTBDDs. Section 6 presents our empirical results. Section 7 concludes the paper.

---

[1] `www.cs.bham.ac.uk/~dxp/prism`

2   The Discrete-Time Setting

DTMCs. Let AP be a fixed, finite set of atomic propositions. A (labelled) DTMC $\mathcal{D}$ is a tuple $(S, \mathbf{P}, L)$ where S is a finite set of *states*, $\mathbf{P} : S \times S \to [0, 1]$ is a *probability matrix* such that $\sum_{s' \in S} \mathbf{P}(s, s') = 1$ for all $s \in S$, and $L : S \to 2^{AP}$ is a *labelling* function which assigns to each state $s \in S$ the set $L(s)$ of atomic propositions that are valid in s. A path through a DTMC is a sequence[2] of states $\sigma = s_0 \, s_1 \, s_2 \ldots$ with $\mathbf{P}(s_i, s_{i+1}) > 0$ for all i. Let $Path^{\mathcal{D}}$ denote the set of all paths in $\mathcal{D}$. $\sigma[i]$ denotes the $(i+1)$th state of σ, i.e. $\sigma[i] = s_{i+1}$. Let $\Pr_s$ denote the unique probability measure on sets of paths that start in state s [22].

PCTL. Let $a \in AP$, $p \in [0, 1]$, k be a natural (or ∞) and $\bowtie \in \{\leqslant, \geqslant\}$. The syntax of PCTL is:

$$\Phi ::= \mathrm{tt} \;\Big|\; a \;\Big|\; \Phi \wedge \Phi \;\Big|\; \neg\Phi \;\Big|\; \mathcal{P}_{\bowtie p}(\Phi \mathcal{U}^{\leqslant k} \Phi)$$

The other boolean connectives are derived in the usual way. For the sake of simplicity, we do not consider the next state operator in this paper. The standard (i.e. unbounded) until formula is obtained by taking k equal to ∞, i.e. $\Phi \mathcal{U} \Psi = \Phi \mathcal{U}^{\leqslant \infty} \Psi$. The semantics of PCTL is defined by [17]:

$$
\begin{aligned}
s &\models \mathrm{tt} && \text{for all } s \in S & s &\models \Phi \wedge \Psi && \text{iff } s \models \Phi \wedge s \models \Psi \\
s &\models a && \text{iff } a \in L(s) & s &\models \mathcal{P}_{\bowtie p}(\Phi \mathcal{U}^{\leqslant k} \Psi) && \text{iff } Prob^{\mathcal{D}}(s, \Phi \mathcal{U}^{\leqslant k} \Psi) \bowtie p \\
s &\models \neg\Phi && \text{iff } s \not\models \Phi
\end{aligned}
$$

$\mathcal{P}_{\bowtie p}(\Phi \mathcal{U}^{\leqslant k} \Psi)$ asserts that the probability measure of the paths that start in s and that satisfy $\Phi \mathcal{U}^{\leqslant k} \Psi$ meets the bound $\bowtie p$. Here,

$$Prob^{\mathcal{D}}(s, \Phi \mathcal{U}^{\leqslant k} \Psi) = \Pr_s\{ \sigma \in Path^{\mathcal{D}} \mid \sigma \models \Phi \mathcal{U}^{\leqslant k} \Psi \}$$

Formula $\Phi \mathcal{U}^{\leqslant k} \Psi$ asserts that Ψ will be satisfied within k steps and that all preceding states satisfy Φ, i.e.:

$$\sigma \models \Phi \mathcal{U}^{\leqslant k} \Psi \text{ iff } \exists j \leqslant k. \, (\sigma[j] \models \Psi \wedge \forall i < j. \, \sigma[i] \models \Phi)$$

Model checking PCTL. PCTL model checking [17] is carried out in the same way as verifying CTL [11] by recursively computing the set $Sat(\Phi) = \{ s \in S \mid s \models \Phi \}$. Checking bounded until formulas amounts to computing the least solution of the following set of equations: $Prob^{\mathcal{D}}(s, \Phi \mathcal{U}^{\leqslant k} \Psi)$ equals 1 if $s \in Sat(\Psi)$,

$$Prob^{\mathcal{D}}(s, \Phi \mathcal{U}^{\leqslant k} \Psi) = \sum_{s' \in S} \mathbf{P}(s, s') \cdot Prob^{\mathcal{D}}(s', \Phi \mathcal{U}^{\leqslant k-1} \Psi) \tag{1}$$

if $s \in Sat(\Phi \wedge \neg\Psi)$ and $k > 0$, and equals 0 otherwise. For DTMC $\mathcal{D} = (S, \mathbf{P}, L)$ and PCTL formula Φ, let DTMC $\mathcal{D}[\Phi] = (S, \mathbf{P}', L)$ where if $s \not\models \Phi$, then $\mathbf{P}'(s, s') = \mathbf{P}(s, s')$ for all $s' \in S$, and if $s \models \Phi$, then $\mathbf{P}'(s, s) = 1$ and $\mathbf{P}'(s, s') = 0$

---

[2] In this paper, we do not dwell upon distinguishing finite and infinite paths.

for all $s' \neq s$. We have $\mathcal{D}[\varPhi][\varPsi] = \mathcal{D}[\varPhi \vee \varPsi]$. Let $\pi^{\mathcal{D}}(s,k)(s')$ denote the probability of being in state s' after k steps in DTMC $\mathcal{D}$ when starting in s, i.e. $\pi^{\mathcal{D}}(s,k)(s') = \Pr_s\{\sigma \in \mathit{Path}^{\mathcal{D}} \mid \sigma[k] = s'\}$.

Proposition 1. *For DTMC $\mathcal{D}$:* $\mathit{Prob}^{\mathcal{D}}(s, \varPhi \mathcal{U}^{\leqslant k} \varPsi) = \sum_{s' \models \varPsi} \pi^{\mathcal{D}[\neg \varPhi \vee \varPsi]}(s,k)(s')$.

Note that $\mathcal{D}[\neg \varPhi \vee \varPsi] = \mathcal{D}[\neg(\varPhi \wedge \varPsi)][\varPsi]$, i.e. all $\neg(\varPhi \wedge \varPsi)$-states and all $\varPsi$-states in $\mathcal{D}$ are made absorbing[3]. The former is correct since $\varPhi \mathcal{U}^{\leqslant k} \varPsi$ is violated as soon as some state is visited that neither satisfies $\varPhi$ nor $\varPsi$. The latter is correct since, once a $\varPsi$-state in $\mathcal{D}$ has been reached (along a $\varPhi$-path) in at most k steps, then $\varPhi \mathcal{U}^{\leqslant k} \varPsi$ holds, regardless of which states will be visited later on.

Model checking $\mathcal{U}^{\leqslant k}$ for all states thus amounts to computing $(\mathbf{P}^{\mathcal{D}[\neg \varPhi \vee \varPsi]})^k \cdot \underline{\iota}_\varPsi$, where $\underline{\iota}_\varPsi$ characterises $\mathit{Sat}(\varPsi)$, i.e. $\iota_\varPsi(s) = 1$ if $s \models \varPsi$, and 0 otherwise. As iterative squaring is not attractive for stochastic matrices due to fill in [28], the product is typically computed in an iterative fashion: $\mathbf{P} \cdot (\ldots (\mathbf{P} \cdot \underline{\iota}_\varPsi)))$.

3   The Continuous-Time Setting

CTMCs. A (labelled) CTMC $\mathcal{C}$ is a tuple $(S, \mathbf{R}, L)$ where S and L are as for DTMCs, and $\mathbf{R} : S \times S \to \mathbb{R}_{\geqslant 0}$ is the *rate matrix*. (We adopt the same conventions as in [6,7], i.e. we do allow self-loops.) The exit rate $E(s) = \sum_{s' \in S} \mathbf{R}(s,s')$ denotes that the probability of taking a transition from s within t time units equals $1 - e^{-E(s) \cdot t}$. If $\mathbf{R}(s,s') > 0$ for more than one state s', a *race* between the outgoing transitions from s exists. That is, the probability $\mathbf{P}(s,s')$ of moving from s to s' in a single step equals the probability that the delay of going from s to s' "finishes before" the delays of any other outgoing transition from s.

Definition 1. *For CTMC $\mathcal{C} = (S, \mathbf{R}, L)$, the* embedded *DTMC is given by* $\mathit{emb}(\mathcal{C}) = (S, \mathbf{P}, L)$, *where* $\mathbf{P}(s,s') = \mathbf{R}(s,s')/E(s)$ *if* $E(s) > 0$, *and* $\mathbf{P}(s,s) = 1$ *and* $\mathbf{P}(s,s') = 0$ *for* $s \neq s'$ *if* $E(s) = 0$.

A path through a CTMC is an alternating sequence $\sigma = s_0 t_0 s_1 t_1 s_2 \ldots$ with $\mathbf{R}(s_i, s_{i+1}) > 0$ and $t_i \in \mathbb{R}_{>0}$ for all i. The time stamps t_i denote the amount of time spent in state s_i. Let $\mathit{Path}^{\mathcal{C}}$ denote the set of paths through $\mathcal{C}$. $\sigma@t$ denotes the state of σ occupied at time t, i.e. $\sigma@t = \sigma[i]$ with i the smallest index such that $t \leqslant \sum_{j=0}^{i} t_j$. Let $\Pr_s$ denote the unique probability measure on sets of paths that start in s [7].

CSL. Let a, p and $\bowtie$ be as before and $t \in \mathbb{R}_{\geqslant 0}$ (or ∞). The syntax of CSL is:

$$\varPhi ::= \mathtt{tt} \;\Big|\; a \;\Big|\; \varPhi \wedge \varPhi \;\Big|\; \neg \varPhi \;\Big|\; \mathcal{S}_{\bowtie p}(\varPhi) \;\Big|\; \mathcal{P}_{\bowtie p}(\varPhi \mathcal{U}^{\leqslant t} \varPhi)$$

$\mathcal{S}_{\bowtie p}(\varPhi)$ asserts that the steady-state probability for a $\varPhi$-state meets the bound $\bowtie p$. The semantics of CSL for the boolean operators is identical to that for

---

[3] That is, the only transitions available in these states are self-loops.

PCTL. For the remaining state formulas [7]:

$$s \models S_{\bowtie p}(\Phi) \qquad \text{iff } \lim_{t \to \infty} \Pr_s \{ \sigma \in Path^{\mathcal{C}} \mid \sigma@t \models \Phi \} \bowtie p$$
$$s \models \mathcal{P}_{\bowtie p}(\Phi \mathcal{U}^{\leqslant t} \Phi) \text{ iff } Prob^{\mathcal{C}}(s, \Phi \mathcal{U}^{\leqslant t} \Phi) \bowtie p$$

The limit in the first equation always exists as $\mathcal{C}$ contains finitely many states [28]. $Prob^{\mathcal{C}}(\cdot)$ is defined in a similar way as for PCTL:

$$Prob^{\mathcal{C}}(s, \Phi \mathcal{U}^{\leqslant t} \Phi) = \Pr_s \{ \sigma \in Path^{\mathcal{C}} \mid \sigma \models \Phi \mathcal{U}^{\leqslant t} \Phi \}.$$

The operator $\mathcal{U}^{\leqslant t}$ is the real-time variant of the PCTL operator $\mathcal{U}^{\leqslant k}$; $\Phi \mathcal{U}^{\leqslant t} \Psi$ asserts that Ψ will be satisfied at some time instant in the interval $[0, t]$ and that at all preceding time instants Φ holds:

$$\sigma \models \Phi \mathcal{U}^{\leqslant t} \Psi \text{ iff } \exists x \leqslant t. \, (\sigma@x \models \Psi \wedge \forall y < x. \, \sigma@y \models \Phi) \,.$$

Note that the standard until operator is obtained by taking t equal to ∞.

CSL model checking [7,6] is performed in the same way as for CTL [11] and PCTL [17], by recursively computing the set $Sat(\Phi)$. For the boolean operators this is exactly as for CTL and for unbounded until this is exactly as for PCTL.

Model checking the S operator. For determining $Sat(S_{\bowtie p}(\Phi))$, first $Sat(\Phi)$ is computed (as usual), and a graph analysis is carried out to determine the bottom strongly connected components (BSCCs) of $\mathcal{C}$, i.e. the set of SCCs in $\mathcal{C}$ that, once entered, cannot be left any more. The steady-state probability distribution π^B inside each BSCC B is determined using standard means [28]: by solving a linear equation system in the size of the BSCC at hand. Then, the probabilities of reaching a BSCC B from a given state s are computed for each B. State s now satisfies $S_{\bowtie p}(\Phi)$ if:

$$\sum_B \left(\Pr\{ \text{reach } B \text{ from } s \} \cdot \sum_{s' \in B \cap Sat(\Phi)} \pi^B(s') \right) \bowtie p$$

All these steps can be performed on the embedded DTMC as timing issues are not involved; for details see [7].

Model checking the $\mathcal{U}^{\leqslant t}$ operator. Checking time-bounded until formulas is based on determining the least solution of the following set of integral equations: $Prob^{\mathcal{C}}(s, \Phi \mathcal{U}^{\leqslant t} \Psi)$ equals 1 if $s \in Sat(\Psi)$,

$$Prob^{\mathcal{C}}(s, \Phi \mathcal{U}^{\leqslant t} \Psi) = \int_0^t \sum_{s' \in S} \mathbf{P}(s, s') \cdot E(s) \cdot e^{-E(s) \cdot x} \cdot Prob^{\mathcal{C}}(s', \Phi \mathcal{U}^{\leqslant t - x} \Psi) \, dx$$

if $s \in Sat(\Phi \wedge \neg \Psi)$, and equals 0 otherwise. Here, $E(s) \cdot e^{-E(s) \cdot x}$ denotes the probability density of taking some outgoing transition from s at time x. Note the resemblance with equation (1) for the PCTL bounded until operator. For

CTMC $\mathcal{C} = (S, \mathbf{R}, L)$ and CSL formula Φ let CTMC $\mathcal{C}[\Phi] = (S, \mathbf{R}', L)$ with $\mathbf{R}'(s, s') = \mathbf{R}(s, s')$ if $s \not\models \Phi$ and 0 otherwise. Note that $emb(\mathcal{C}[\Phi]) = emb(\mathcal{C})[\Phi]$. It has been shown in [6] that for a given CTMC $\mathcal{C}$ and state s in $\mathcal{C}$, the measure $Prob^{\mathcal{C}}(s, \Phi \mathcal{U}^{\leqslant t} \Psi)$ can be calculated by means of a transient analysis of the CTMC $\mathcal{C}'$, which can easily be derived from $\mathcal{C}$ using the $[\cdot]$ operator. Let $\pi^{\mathcal{C}}(s, t)(s')$ denote the probability of being in state s' at time t given that the system started in state s, i.e. $\pi^{\mathcal{C}}(s, t)(s') = \Pr_s\{\sigma \in Path^{\mathcal{C}} \mid \sigma@t = s'\}$.

Theorem 1. *[6] For CTMC $\mathcal{C}$:* $Prob^{\mathcal{C}}(s, \Phi \mathcal{U}^{\leqslant t} \Psi) = \sum_{s' \models \Psi} \pi^{\mathcal{C}[\neg \Phi \vee \Psi]}(s, t)(s')$.

4 Faster Time-Bounded until Verification

In this section, we present an algorithm for verifying time-bounded until formulas that is based on (i) the aforementioned reduction to transient analysis and on (ii) the algorithm for PCTL bounded until. This combination – suggested by the strong resemblance of Theorem 1 and Proposition 1 – yields an improvement of $\mathcal{O}(N)$ in time complexity over the algorithm suggested in [6], where N is the number of states in the CTMC. We first briefly describe uniformisation.

Uniformisation. Uniformisation is a transformation of a CTMC into a DTMC:

Definition 2. *For CTMC $\mathcal{C} = (S, \mathbf{R}, L)$ the* uniformised *DTMC is given by* $unif(\mathcal{C}) = (S, \mathbf{P}, L)$ *where* $\mathbf{P} = \mathbf{I} + \mathbf{Q}/q$ *for* $q \geqslant \max\{E(s) \mid s \in S\}$ *and* $\mathbf{Q} = \mathbf{R} - diag(\underline{E})$.

The *uniformisation rate* q is determined by the state with the shortest mean residence time. All (exponential) delays in the CTMC $\mathcal{C}$ are normalised with respect to q. That is, for each state $s \in S$ with $E(s) = q$, one epoch in $unif(\mathcal{C})$ corresponds to a single exponentially distributed delay with rate q, after which one of its successor states is selected probabilistically. As a result, such states have no self-loop in the DTMC. If $E(s) < q$ – this state has on average a longer state residence time than $\frac{1}{q}$ – one epoch in $unif(\mathcal{C})$ might not be "long enough". Hence, in the next epoch these states might be revisited and, accordingly, are equipped with a self-loop with probability $1 - \frac{E(s)}{q}$. Note the difference between the embedded DTMC $emb(\mathcal{C})$ and the uniformised DTMC $unif(\mathcal{C})$: whereas the epochs in $\mathcal{C}$ and $emb(\mathcal{C})$ coincide and $emb(\mathcal{C})$ can be considered as the timeless variant of $\mathcal{C}$, a single epoch in $unif(\mathcal{C})$ corresponds to a single exponentially distributed delay with rate q in $\mathcal{C}$.

Transient analysis. The probabilities $\pi^{\mathcal{C}}(s, t)(s')$ are now computed as follows:

$$\underline{\pi}(s, t) = \underline{\pi}(s, 0) \cdot \sum_{k=0}^{\infty} e^{-q \cdot t} \frac{(q \cdot t)^k}{k!} \mathbf{P}^k = \sum_{k=0}^{\infty} \gamma(k, q \cdot t) \cdot \underline{\pi}(s, k) \qquad (2)$$

where $\mathbf{P}$ is the probability matrix of the DTMC $unif(\mathcal{C})$, and $\gamma(k, q \cdot t)$ is the kth Poisson probability with parameter $q \cdot t$, i.e. $\gamma(k, q \cdot t) = e^{-q \cdot t} \cdot (q \cdot t)^k / k!$. The vector $\pi(s,k)$ denotes the probability distribution in $unif(C)$ after k epochs when starting in s, i.e. $\pi(s,k) = \pi(s,0) \cdot \mathbf{P}^k$, where $\pi(s,0)(s) = 1$ and $\pi(s,0)(s') = 0$ if $s \neq s'$. Equation (2) can be understood as follows. During the time interval $[0,t)$, with probability $\gamma(k,q\cdot t)$ exactly k jumps have taken place in the DTMC $unif(C)$. The effect of these jumps is described by $\pi(s,0)\cdot\mathbf{P}^k$. Weighting this vector with $\gamma(k,q\cdot t)$ and summing over all possible numbers of jumps in $[0,t)$, we obtain, by the law of total probability, the probability vector $\pi(s,t)$.

Given an accuracy ϵ, the number of terms R_ϵ of the infinite summation in (2) that have to be considered is the smallest value satisfying:

$$\sum_{n=0}^{R_\epsilon} \frac{(q\cdot t)^n}{n!} \geq \frac{1-\epsilon}{e^{-q\cdot t}} = (1-\epsilon)\cdot e^{q\cdot t}$$

For large $q\cdot t$, R_ϵ is of order $O(q\cdot t)$.[4] As the first group of Poisson probabilities are typically very small, the first L_ϵ terms in (2) are negligible and need not be computed. L_ϵ and R_ϵ are called the left and right truncation point, respectively.

A first algorithm. The algorithm for time-bounded until as suggested in [6] is based on carrying out a computation according to equation (2) and the fact that $\pi(s,k) = \pi(s,0) \cdot \mathbf{P}^k$. The computation is carried out in an iterative manner per individual state s starting from the initial distribution $\pi(s,0)$. The pseudo-code of this algorithm is presented in Fig. 1. Here, and in the subsequent algorithms in this paper, the Poisson probabilities are computed using the Fox-Glynn algorithm [15] that avoids overflow for large $q\cdot t$. The overall time complexity of this procedure is $\mathcal{O}(N\cdot q\cdot t\cdot M)$, where q is the uniformisation rate of the CTMC at hand, t the time bound of the until formula, N the number of states and M the number of non-zero entries in $\mathbf{R}$. This follows directly from the fact that for each state the number of terms of (2) that needs to be considered is $\mathcal{O}(q\cdot t)$, where each term requires a matrix vector multiplication with $\mathcal{O}(M)$ multiplications given a sparse data structure.

An alternative algorithm. The basic idea of the new algorithm is to use the iterative matrix vector multiplication of the PCTL bounded until operator as a basis, and impose the computation of the Poisson probabilities on top of it. This is suggested by the following observation:

Proposition 2. $Prob^C(s, \Phi \mathcal{U}^{\leqslant t} \Psi) = \sum_{k=0}^{\infty} \gamma(k,q\cdot t) \cdot Prob^{unif(C)}(s, \Phi \mathcal{U}^{\leqslant k} \Psi)$

Recall that $\gamma(k,q\cdot t)$ denotes the probability of taking k jumps in the DTMC $unif(C)$ in the interval $[0,t)$. From Propositions 1 and 2 it follows that the vector $\underline{Prob}^C(\Phi \mathcal{U}^{\leqslant t} \Psi)$ can be obtained in an iterative manner, cf. the pseudo-code in Fig. 2. As a result, a global transient analysis is carried out, yielding for each state s the probability measure $Prob^C(s, \Phi \mathcal{U}^{\leqslant t} \Psi)$. Note that, as opposed to the

---

[4] Note that the DTMC $unif(C)$ may reach steady state before R_ϵ and, in this case, the summation can be truncated at this earlier point [24].

```
// compute Poisson probabilities
γ, L_ε, R_ε := FoxGlynn(q · t, ε)
// main loop
foreach s ∈ S
    sol := 0
    p := π(s, 0)
    for k = 1 to L_ε − 1
        p := p · P
    endfor
    for k = L_ε to R_ε
        p := p · P
        sol := sol + γ(k, q·t) · p
    endfor
    // Prob(s, Φ U^{⩽t} Ψ) = sol · ι_Ψ^T
endfor
```

Fig. 1. A first algorithm

```
// compute Poisson probabilities
γ, L_ε, R_ε := FoxGlynn(q · t, ε)
// main loop
sol := 0
b := ι_Ψ
for k = 1 to L_ε − 1
    b := P · b
endfor
for k = L_ε to R_ε
    b := P · b
    sol := sol + γ(k, q·t) · b
endfor
// Prob(Φ U^{⩽t} Ψ) = sol
```

Fig. 2. An efficient variant

algorithm of Fig. 1, sol is not a probability vector in Fig. 2, i.e. its elements do not sum up to one. It is evident from the efficiency considerations given just before, that the time complexity of the adapted algorithm is $\mathcal{O}(q{\cdot}t{\cdot}M)$, thus yielding an improvement of $\mathcal{O}(N)$ over the previous algorithm.

5   Symbolic Model Checking CTMCs with PRISM

Due to the recent improvements in verification time – including our suggested improvement – space efficiency considerations become more important for CTMC model checking. In this section, we report on symbolic model checking of CTMCs (against CSL) using MTBDDs. MTBDDs have the ability to exploit structure (regularity) in models and can represent them in a far more compact way than a sparse matrix would. The success of BDD-based model checking in the non-probabilistic case serves as sufficient motivation to develop the foundations of MTBDD-based model checking and experiment with these techniques.

MTBDDs. Let $x_1 < x_2 < \ldots < x_n$ be distinct, totally ordered state variables. An MTBDD over $(x_1, \ldots, x_n)$ is a rooted directed graph with vertex set V containing two types of vertices:

- each *non-terminal vertex* v is labelled by a state variable $var(v) \in \{x_1, \ldots, x_n\}$ and has two children $left(v), right(v) \in V$
- each *terminal vertex* v is labelled by a real number $val(v)$,

such that $var(v) < var(w)$ for each non-terminal vertex v and non-terminal child w of v. The constraint requires that on any path from the root to a terminal vertex, the variables respect the ordering $<$. An MTBDD M over $(x_1, \ldots, x_n)$ represents the function $f_M : \{0, 1\}^n \to \mathbb{R}$, whose values are obtained by traversing M starting at the root vertex as follows. For non-terminal vertex v, the edge from v to $left(v)$ represents the case when $var(v)$ is false; the edge from v to $right(v)$ the case $var(v)$ is true. For efficiency reasons, MTBDDs are usually stored in a reduced form [10]. Note that a BDD is an MTBDD with $val(v) \in \{0,1\}$ for all terminal vertices v.

Representing CTMCs by MTBDDs. Let $\mathcal{C} = (S, \mathbf{R}, L)$ be a CTMC with $|S| = 2^n$ and L injective. (Any labelled CTMC may be transformed into one satisfying these conditions by adding dummy states and new propositions.) Let $a_1, \ldots, a_n$ be an enumeration of the atomic propositions and identify each state s with the boolean n-tuple $(b_1, \ldots, b_n)$ where $b_i = 1$ iff $a_i \in L(s)$. This encoding of states is standard [5,7,13]. Thus, $S = \{0,1\}^n$ where each state s is identified with its encoding and $\mathbf{R}$ with the function $F : \{0,1\}^{2n} \to \mathbb{R}$ where $F(x_1, y_1, \ldots, x_n, y_n) = \mathbf{R}((x_1, \ldots, x_n), (y_1, \ldots, y_n))$.

Operations on MTBDDs. Model checking CTMCs can be performed with standard operations on MTBDDs. For completeness, we briefly describe these here. The operator APPLY allows a point-wise application of the binary operator op (e.g. $+$ or $\times$) to two MTBDDs. For MTBDDs M_1 and M_2, APPLY$(op, \mathsf{M}_1, \mathsf{M}_2)$ yields an MTBDD for function $f_{\mathsf{M}_1} \, op \, f_{\mathsf{M}_2}$. For MTBDDs R and b representing matrix $\mathbf{R}$ and vector $\underline{b}$ respectively, MTBDD MVMULT(R, b) represents the vector $\mathbf{R} \cdot \underline{b}$. For $q \in \mathbb{R}$, CONST(q) denotes the MTBDD consisting of a single terminal vertex v with $val(v) = q$. For an MTBDD M, FINDMAX(M) returns the maximum value of the terminal vertices of M. The COMP operator takes an MTBDD M and an interval $I \subseteq \mathbb{R}$ and returns the BDD representing the function that equals 1 if $f_{\mathsf{M}}(x_1, \ldots, x_n) \in I$ and 0 otherwise. Operator ABSTRACT$(op, \mathsf{M}, x_1, \ldots, x_n)$ returns the MTBDD which results from abstracting all of the variables $x_1, \ldots, x_n$ from M by applying op over all possible values taken by these variables.

MTBDD-based model checking of CSL. The symbolic model checking algorithm for CSL is identical to the one proposed in [7] except that we use transient analysis and uniformisation rather than numerical integration (which needs dedicated variants of MTBDDs). Let $\mathcal{C} = (S, \mathbf{R}, L)$ be a CTMC represented by MTBDD R as explained above. For each CSL formula Φ a BDD Sat$[\![\Phi]\!]$ is defined that represents the characteristic function of the set $Sat(\Phi)$. By applying standard operators on MTBDDs we determine the MTBDDs P representing the transition probability matrix $\mathbf{P}$ of $emb(\mathcal{C})$, and E the vector of exit rates $\underline{E}$. Then:

$$\mathrm{Sat}[\![\, \mathrm{tt} \,]\!] = \mathrm{CONST}(1)$$
$$\mathrm{Sat}[\![\, a_i \,]\!] = \text{the BDD for the boolean function } (x_1, \ldots, x_n) \mapsto x_i$$
$$\mathrm{Sat}[\![\, \neg\Phi \,]\!] = \mathrm{NOT}(\mathrm{Sat}[\![\Phi]\!])$$
$$\mathrm{Sat}[\![\, \Phi \wedge \Psi \,]\!] = \mathrm{AND}(\mathrm{Sat}[\![\Phi]\!], \mathrm{Sat}[\![\Psi]\!])$$
$$\mathrm{Sat}[\![\, \mathcal{S}_{\bowtie p}(\Phi) \,]\!] = \mathrm{COMP}(\mathrm{STEADYSTATE}(\mathsf{P}, \mathrm{Sat}[\![\Phi]\!]), \bowtie p)$$
$$\mathrm{Sat}[\![\, \mathcal{P}_{\bowtie p}(\Phi \mathcal{U} \Psi) \,]\!] = \mathrm{COMP}(\mathrm{UNTIL}(\mathsf{P}, \mathrm{Sat}[\![\Phi]\!], \mathrm{Sat}[\![\Psi]\!]), \bowtie p)$$
$$\mathrm{Sat}[\![\, \mathcal{P}_{\bowtie p}(\Phi \mathcal{U}^{\leqslant t} \Psi) \,]\!] = \mathrm{COMP}(\mathrm{TBUNTIL}(\mathsf{R}, \mathsf{E}, \mathrm{Sat}[\![\Phi]\!], \mathrm{Sat}[\![\Psi]\!], t, \epsilon), \bowtie p).$$

```
algorithm TBUNTIL(R, E, Sat[[Φ]], Sat[[Ψ]], t, ε)
    // uniformisation
    R' := APPLY(×, NOT(OR(NOT(Sat[[Φ]]), Sat[[Ψ]])), R)
    E' := APPLY(×, NOT(OR(NOT(Sat[[Φ]]), Sat[[Ψ]])), E)
    q := FINDMAX(E')
    Q := APPLY(−, R', APPLY(×, E', Identity))
    P := APPLY(+, I, APPLY(÷, Q, CONST(q)))
    // compute Poisson probabilities
    γ, Lε, Rε := FOXGLYNN(q · t, ε)
    // main loop
    sol := CONST(0)
    b := Sat[[Ψ]]
    for k = 1 to Lε − 1
        b := MVMULT(P, b)
    endfor
    for k = Lε to Rε
        b := MVMULT(P, b)
        sol := APPLY(+, sol, APPLY(×, CONST(γ(k, q·t)), b))
    endfor
    return sol
end.
```

Fig. 3. MTBDD algorithm for CSL time-bounded until using transient analysis

Here, $Sat[[a_i]]$ is a BDD consisting of a single state vertex v labelled with x_i such that $left(v)$ and $right(v)$ are labelled with 0 and 1, respectively. The steady state and unbounded until operators are treated symbolically as described in [5] and [7] respectively.

TBUNTIL assigns to each state $s \in S$ the probability (with precision ϵ) of the set of paths that start in s fulfilling $\Phi \mathcal{U}^{\leqslant t} \Psi$, i.e. it represents the function $s \mapsto Prob(s, \Phi \mathcal{U}^{\leqslant t} \Psi)$. The algorithm for TBUNTIL is shown in Fig. 3, where Identity denotes the MTBDD representing an identity matrix of the appropriate size. In the first two lines, the CTMC $\mathcal{C}[\neg \Phi \vee \Psi]$ is computed. Note that the APPLY operator filters out the states that become absorbing, i.e. the states that do not satisfy $\neg(\neg \Phi \vee \Psi)$. In the subsequent three lines, the uniformisation rate q and the DTMC $unif(\mathcal{C})$ are determined. The rest of the pseudo-code is the MTBDD-based counterpart of the algorithm shown earlier in Fig. 2.

PRISM. PRISM is a verifier for discrete probabilistic systems such as DTMCs (against PCTL) and MDPs (against PCTL [9] with fairness [8]). The tool is implemented using a combination of Java and C++. The high level parts of the tool, such as the user interface and the parsers, are written in Java. The engines and libraries are mostly written in C++. PRISM takes as input a model description in a probabilistic variant of reactive modules [1], constructs the model from its description and computes the set of reachable states. Model checking

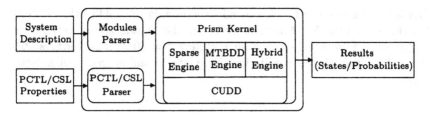

Fig. 4. The PRISM tool architecture

using different data structures is supported, cf. Fig. 4: symbolic representations using (MT)BDDs, conventional sparse matrices, and a hybrid approach using MTBDDs for storing matrices and conventional representations for probability vectors. For the manipulation of the symbolic data structures, PRISM uses the CUDD package [27] which is written in C. More information about the tool can be found at www.cs.bham.ac.uk/~dxp/prism.

The MTBDD based model checking algorithm for CSL has been implemented in PRISM, thus extending its applicability to *continuous* probabilistic systems. The realisation in PRISM includes an "on-the-fly" steady state detection as part of the transient analysis (as in [19]). For the sake of clarity, this mechanism is not included in the algorithm in Fig. 3.

6   Experiments

The case studies. To facilitate a comparison with $E \vdash MC^2$ [19], we consider two case studies that have been verified previously with CSL: a tandem network [20] and a cyclic server polling system [21].

The tandem network consists of a $M/Cox_2/1$-queue and a $M/M/1$-queue, both of capacity K, put in sequence. Jobs arrive at the first station with rate $4 \cdot K$. The first server executes jobs in either done or two phases, i.e. with probability 0.9 a job is served once (with rate 2), and with probability 0.1 the job has to pass an additional phase (with rate 2). Once served by the first station, jobs are queued in the second station where service takes place with rate 4. The properties we verify for this model are the following probabilistic path properties:

- $\Diamond^{\leq t}$ *full*, i.e. the tandem network becomes fully occupied within t time units
- $\Diamond^{\leq t}$ *fst*, i.e. the first station of the tandem network becomes fully occupied within t time units

where $\Diamond^{\leq t} \Phi \equiv tt\, \mathcal{U}^{\leq t} \Phi$.

The polling server [21] polls K stations in a cyclic fashion. The times for generating a message, for polling a station and for serving a job by a station are all distributed exponentially with rates $1/K$, 200 and 1, respectively. If the server finds a station idle, then the service time is zero. For this system, we check the property $busy_1 \Rightarrow \mathcal{P}_{\bowtie p}(\Diamond^{\leq t} poll_1)$, i.e. once the first station has a job to be served it will be polled within t time units with probability $\bowtie p$.

Statistics and assessment. We ran all experiments on a 440 MHz SUN Ultra 10 workstation with 512 Mb memory under the Solaris 2.7 operating system. All properties were checked with an accuracy $\varepsilon = 10^{-6}$. The verifiers PRISM and $E \vdash MC^2$ provide the results for the symbolic and sparse implementations respectively.

Table 1. Comparison of the original and improved sparse implementations of time-bounded until algorithm

model	K	# states	formula	time (in *sec*) original	improved
tandem network	2	15	$\mathcal{P}_{\bowtie p}(\lozenge^{\leqslant 2} full)$	0.07	0.01
			$\mathcal{P}_{\bowtie p}(\lozenge^{\leqslant 10} full)$	0.13	0.01
			$\mathcal{P}_{\bowtie p}(\lozenge^{\leqslant 100} full)$	0.71	0.03
			$\mathcal{P}_{\bowtie p}(\lozenge^{\leqslant 1000} full)$	1.29	0.09
	20	861	$\mathcal{P}_{\bowtie p}(\lozenge^{\leqslant 2} full)$	563.94	0.55
			$\mathcal{P}_{\bowtie p}(\lozenge^{\leqslant 10} full)$	1927.59	1.01
			$\mathcal{P}_{\bowtie p}(\lozenge^{\leqslant 100} full)$	1978.22	1.05
			$\mathcal{P}_{\bowtie p}(\lozenge^{\leqslant 1000} full)$	1954.01	1.00
polling system	3	36	$busy_1 \Rightarrow \mathcal{P}_{\bowtie p}(\lozenge^{\leqslant 2} poll_1)$	0.96	0.02
	5	240	$busy_1 \Rightarrow \mathcal{P}_{\bowtie p}(\lozenge^{\leqslant 2} poll_1)$	59.55	0.16
	7	1,344	$busy_1 \Rightarrow \mathcal{P}_{\bowtie p}(\lozenge^{\leqslant 2} poll_1)$	2637.11	1.71
	10	15,360	$busy_1 \Rightarrow \mathcal{P}_{\bowtie p}(\lozenge^{\leqslant 2} poll_1)$	–	13.48

Statistics and assessment for the improved method. The statistics in Table 1 compare the verification times of the original sparse implementation of time-bounded until (Fig. 1) and the improved algorithm (Fig. 2). As expected, the results confirm that the improved algorithm is a factor of N faster, where N is the size of the state space. We note that, in several cases, there is an even greater speed-up. This is possibly due to the different computation steps performed by the two algorithms: the original works via a forwards exploration of the state space, whereas the improved version works backwards. To see this, note the difference between the iteration steps $\underline{p} := \underline{p} \cdot \mathbf{P}$ in the original as opposed to $\underline{b} := \mathbf{P} \cdot \underline{b}$ in the improved.

Statistics and assessment for the symbolic implementation. We now compare our symbolic implementation of time-bounded until with its sparse counterpart. For the tandem network and polling system examples, we have constructed efficient MTBDD representations of the transition matrix using the methods presented in [14] (for further details see www.cs.bham.ac.uk/~dxp/prism). This allows us to build and store much larger models with MTBDDs (given regularity) than is feasible with a sparse implementation.

Table 2. Comparison of symbolic and sparse verification of the tandem network

sparse versus symbolic implementation					
K	# states	\multicolumn{4}{c}{time per iteration (in *sec*)}			
		$\mathcal{P}_{\bowtie p}(\Diamond^{\leq 2} \textit{full})$		$\mathcal{P}_{\bowtie p}(\Diamond^{\leq 2} \textit{fst})$	
		symbolic	sparse	symbolic	sparse
63	8,128	0.08	0.01	0.02	0.01
127	32,640	0.17	0.04	0.04	0.05
255	130,816	0.37	0.55	0.06	0.15
511	523,776	0.81	1.50	0.10	0.71
1023	2,096,128	–	–	0.23	–
2047	8,386,560	–	–	0.31	–
4095	33,550,336	–	–	0.66	–

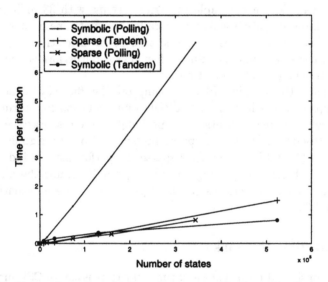

Fig. 5. Comparison of symbolic and sparse verification of both examples

The results of the comparison of our symbolic implementation with the sparse implementation are presented in Table 2 and Fig. 5. We have measured time per iteration as both implementations follow the improved algorithm given in Fig. 2. Table 2 summarises the results for the tandem network example based on two CSL properties. Fig. 5 gives time per iteration plotted against the size of the state space for both the tandem network and polling system for the CSL properties $\mathcal{P}_{\bowtie p}(\Diamond^{\leq 2} \textit{full})$ and $busy_1 \Rightarrow \mathcal{P}_{\bowtie p}(\Diamond^{\leq 2} poll_1)$ respectively.

The efficiency of the symbolic time-bounded until implementation depends on the size of the MTBDDs representing the iteration vectors ($\underline{b}$ and $\underline{sol}$ in Fig. 2). Our experiments show that these are usually significantly larger than the MTBDD for the transition matrix, because of their relative lack of structure. For vectors to be represented compactly by MTBDDs the main requirement is a limited number of distinct elements. This condition is dependent on both the structure of the model and on the property being verified, and as such it is difficult to determine when these vectors will be represented compactly. For example, compare the difference in performance of the symbolic implementation on two different models, as shown in Fig. 5. The times for the tandem network are much faster than for the polling system for models of equivalent size.

On the other hand, in the sparse implementation the time complexity is dependent purely on the number of non-zeros in the matrix used for the computation. In Fig. 5, the times for the sparse approach can be seen to be almost identical for the tandem network and the polling system (note that in both examples the number of non-zeros in the rate matrix is linear in the size of the state space).

Comparing the results for the two implementations confirms, as expected, that we can verify larger models using a symbolic as opposed to a sparse approach; for example, we were able to verify systems with 33 million states. A more surprising observation which we note for the first time is that, for certain models and certain properties, symbolic analysis is faster than sparse. So far, see e.g. [14], the sparse implementation has always outperformed the MTBDDs on quantitative numerical calculations.

We are currently extending PRISM to improve the efficiency further by taking a *hybrid* approach which uses an MTBDD representation for storing matrices and a conventional representation for probability vectors. Early experiments show that, although slower than a sparse implementation, it is significantly faster than the pure MTBDD version. Like sparse, its performance is independent of the regularity of the model being considered, but it retains the advantage of MTBDDs, in that larger models can be represented. More information will be available in [26].

7   Concluding Remarks

This paper considered both space and time efficiency issues of CSL model checking of CTMCs. We presented an improvement in time efficiency of $\mathcal{O}(N)$ for verifying time-bounded until formulas. The obtained empirical results indicate a drastic improvement in run times, making model checking of systems of realistic size feasible. In addition, we reported on symbolic model checking of CSL using MTBDD based uniformisation and transient analysis. Although, for simplicity, we have restricted the exposition in this paper to an until operator with time bounds $[0, t]$, the results of our paper carry over to $\mathcal{U}^I$ for arbitrary interval $I \subseteq \mathbb{R}_{\geqslant 0}$ in a straightforward manner.

Acknowledgements. Joachim Meyer-Kayser, Hannes Bruchner and Markus Siegle (all of the University of Erlangen-Nürnberg) are kindly acknowledged for adapting the model checker E ⊢ MC^2 to the new algorithm for checking time-bounded until formulas and for providing us with the new version of the tool.

The last three authors are members of the ARC project 1031 "Stochastic Modelling and Verification" funded by the British Council and DAAD.

References

1. R. Alur and T.A. Henzinger. Reactive modules. In *IEEE Symp. on Logic in Computer Science*, 207–218, 1996.
2. A. Aziz, K. Sanwal, V. Singhal and R. Brayton. Verifying continuous time Markov chains. In *Computer-Aided Verification*, LNCS 1102: 269–276, 1996.
3. A. Aziz, K. Sanwal, V. Singhal and R. Brayton. Model checking continuous time Markov chains. *ACM Trans. on Computational Logic*, 1(1): 162–170, 2000.
4. R.I. Bahar, E.A. Frohm, C.M. Gaona, G.D. Hachtel, E. Macii, A. Pardo and F. Somenzi. Algebraic decision diagrams and their applications. *Formal Methods in System Design*, 10(2/3): 171–206, 1997.
5. C. Baier, E. Clarke, V. Hartonas-Garmhausen, M. Kwiatkowska, and M. Ryan. Symbolic model checking for probabilistic processes. In *Automata, Languages and Programming*, LNCS 1256: 430–440, 1997.
6. C. Baier, B.R. Haverkort, H. Hermanns and J.-P. Katoen. Model checking continuous-time Markov chains by transient analysis. In *Computer Aided Verification*, LNCS 1855: 358–372, 2000.
7. C. Baier, J.-P. Katoen and H. Hermanns. Approximate symbolic model checking of continuous-time Markov chains. In *Concurrency Theory*, LNCS 1664: 146–162, 1999.
8. C. Baier and M. Kwiatkowska. Model checking for a probabilistic branching-time logic with fairness. *Distr. Comp.*, 11(3): 125–155, 1998.
9. A. Bianco and L. de Alfaro. Model checking of probabilistic and nondeterministic systems. In *Found. of Softw. Techn. and Th. Comp. Sc.*, LNCS 1026: 499–513, 1995.
10. K. Brace, R. Rudell and R. Bryant. Efficient implementation of a BDD package. In: *27th ACM/IEEE Design Automation Conference*, 1990.
11. E. Clarke, E. Emerson and A. Sistla. Automatic verification of finite-state concurrent systems using temporal logic specifications. *ACM Trans. on Progr. Lang. and Sys.*, 8: 244–263, 1986.
12. E. Clarke, M. Fujita, P.C. McGeer and J.C-Y. Yang. Multi-terminal binary decision diagrams: an efficient data structure for matrix representation. In *Formal Methods in System Design*, 10(2/3): 149–169, 1997.
13. E. Clarke, O. Grumberg and D. Long. Verification tools for finite-state concurrent programs. In *A Decade of Concurrency*, LNCS 803: 124–175, 1993.
14. L. de Alfaro, M. Kwiatkowska, G. Norman, D. Parker and R. Segala. Symbolic model checking for probabilistic processes using MTBDDs and the Kronecker representation. In *Tools and Algorithms for the Analysis and Construction of Systems*, LNCS 1785: 395–410, 2000.
15. B.L. Fox and P.W. Glynn. Computing Poisson probabilities. *Comm. of the ACM*, 31(4): 440–445, 1988.
16. D. Gross and D.R. Miller. The randomization technique as a modeling tool and solution procedure for transient Markov chains. *Oper. Res.* 32(2): 343–361, 1984.
17. H.A. Hansson and B. Jonsson. A logic for reasoning about time and reliability. *Form. Asp. of Comp.*, 6(5): 512–535, 1994.
18. V. Hartonas-Garmhausen, S. Campos and E.M. Clarke. PROBVERUS: probabilistic symbolic model checking. In *Formal Methods for Real-Time and Prob. Sys.*, LNCS 1601: 96–111, 1999.
19. H. Hermanns, J.-P. Katoen, J. Meyer-Kayser and M. Siegle. A Markov chain model checker. In *Tools and Algorithms for the Construction and Analysis of Systems*, LNCS 1785: 347–362, 2000.

20. H. Hermanns, J. Meyer-Kayser and M. Siegle. Multi-terminal binary decision diagrams to represent and analyse continuous-time Markov chains. In *Proc. 3rd Int. Workshop on the Num. Sol. of Markov Chains*, pp. 188-207, 1999.

21. O.C. Ibe and K.S. Trivedi. Stochastic Petri net models of polling systems. *IEEE J. on Sel. Areas in Comms.*, 8(9): 1649–1657, 1990.

22. J. Kemeny, J. Snell and A. Knapp. *Denumerable Markov Chains*. Van Nostrand, 1966.

23. A. Jensen. Markov chains as an aid in the study of Markov processes. *Skand. Aktuarietidskrift*, **3**: 87–91, 1953.

24. J.K. Muppala and K.S. Trivedi. Numerical transient solution of finite Markovian queueing systems. In U. Bhat, ed, *Queueing and Related Models*, Oxford University Press, 1992.

25. W.D. Obal II and W.H. Sanders. State-space support for path-based reward variables. *Perf. Ev.*, **35**: 233–251, 1999.

26. D. Parker. Implementation of symbolic model checking for probabilistic systems. Ph.D thesis, School of Computer Science, University of Birmingham, 2001 (to appear).

27. F. Somenzi. CUDD: CU decision diagram package. Public software, Colorado University, Boulder, 1997.

28. W. Stewart. *Introduction to the Numerical Solution of Markov Chains*. Princeton Univ. Press, 1994.

Reachability Analysis of Probabilistic Systems by Successive Refinements

Pedro R. D'Argenio[1]*, Bertrand Jeannet[2],
Henrik E. Jensen[2], and Kim G. Larsen[2,1]

[1] Faculty of Informatics, University of Twente
P.O. Box 217. NL-7500 AE - Enschede. The Netherlands
dargenio@cs.utwente.nl
[2] BRICS - Aalborg University
Frederik Bajers vej 7-E. DK-9220 Aalborg. Denmark
{bjeannet,ejersbo,kgl}@cs.auc.dk

Abstract. We report on a novel development to model check quantitative reachability properties on Markov decision processes together with its prototype implementation. The innovation of the technique is that the analysis is performed on an abstraction of the model under analysis. Such an abstraction is significantly smaller than the original model and may safely refute or accept the required property. Otherwise, the abstraction is refined and the process repeated. As the numerical analysis necessary to determine the validity of the property is more costly than the refinement process, the technique profits from applying such numerical analysis on smaller state spaces.

1 Introduction

The verification of systems has nowadays reached a clear maturity. Fully automatic tools, in particular model checkers, have been developed and successfully used in industrial cases. A model checker is a tool that can answer whether the system under study satisfies some required property. Many times, however, these type of properties are not expressive enough to assert adequately the correctness of a system. Nevertheless, it is desirable that the probability of reaching the unavoidable error is small enough. *Quantitative model checking*, that is, model checking of probabilistic models with respect to *probabilistic* properties, has already been studied during the last decade [13,2,5,20,4, etc.]. However, it was not until recently that attention was drawn to efficient tool implementations. In this paper we report on a novel development to model check quantitative properties.

We use *Markov decision processes* (see e.g. [27]) to describe the system under study. This model, also called probabilistic transition system (PTS), allows to combine probabilistic and non-deterministic steps and is a natural extension to traditional non-deterministic models (such as labelled transition systems). Our preference for a probabilistic model that allows non-determinism is based on two

---

* Supported by the STW-PROGRESS project TES-4999

L. de Alfaro and S. Gilmore (Eds.): PAPM-PROBMIV 2001, LNCS 2165, pp. 39–56, 2001.

facts. First, PTSs are closed under parallel compostition which facilitates the modelling process. Second, PTSs are also closed under abstraction. This reason is fundamental as the method introduced in this paper is based on abstraction techniques.

We focus on a restricted set of reachability properties. They allow to specify that the probability to reach a particular final condition f from any state satisfying a given initial condition i is smaller (or greater) than a probability p. This type of properties is not so restrictive as it seems since we can always use checking automata to add additional constraints to the property.

The method we present is based on automatic abstraction and refinement techniques. The basic idea is to use abstraction to reduce the high cost of probabilistic analysis. The difficulty lies in finding the right abstraction level, depending on the property to prove. To address it, the method starts with a coarse abstraction of the system which is obtained by *partitioning* the state space, according to the property under study. The property is then checked on the obtained abstract model. The verdict may be inconclusive, that is, p happens to be between the calculated upper bound of the minimum and the lower bound of the maximum actual probabilities. In this case the previous abstraction is refined and the question posed again. The process is successively repeated until a satisfactory answer is given, or no further refinement is possible. To efficiently store the state space, perform abstractions and process the refinement steps, we use BDDs and MTBDDs (more precisely ADDs) [10,3]. The soundness of the method is asserted by considering a suitable probabilistic simulation [22,28] (which preserves the kind of property we consider), and by showing that abstraction by partitioning respects this simulation relation.

The contributions of this paper are first the definition of the probabilistic simulation relation that allows to prove the soundness of our method, and secondly, the design of efficient algorithms to abstract PTSs, to analyse and to refine them. Finally, experimental results shows the effectiveness of the method.

Related work. The partition refinement method we use on PTSs resorts to principles already applied to finite-state systems [6] and timed automata [1]. However our aim is not to generate a minimal model w.r.t. a bisimulation relation, but to steer the refinement process in order to prove as early as possible an intended (probabilistic) property.

The efficiency provided by MTBDDs to store and logically manipulate the state space made them also the choice of recent quantitative model checkers [14, 9]. However, if it comes to model analysis via numerical recipes like simplex or (iterative) solutions of equations systems, experience has shown that MTBDDs do not outperform classical data structures (such as sparse matrices) [3,18,9]. The main reason appears to be that any of these algorithms tend to require the storage of a distinct real number per actual state [16]. In our case, the use of MTBDDs is focus on the manipulation of probabilistic transition relations and its use in the abstraction techniques. After abstraction, the size of the problem submitted to numerical analysis becomes a significantly smaller issue.

Other quantitative model checkers have also been developed. The tool PROB-VERUS [14] allows to check the validity of a PCTL formula [13] on a (discrete time) Markov chain. Therefore, models do not contain non-determinism. Instead, PRISM [9] is a quantitative model checker for PCTL formulas on (discrete time) Markov decision processes, i.e., non-determinism is inherent to the model. Like PRISM, we also do model checking on Markov decision processes, but we restrict to a particular kind of PCTL formula. For completeness reason, we also mention the quantitative model checker $E \vdash MC^2$ [17], which model checks probabilistic timed properties on continuous-time Markov chains.

Organization of the paper. Section 2 and Section 3 introduce the theoretical foundations of the implemented tool. The algorithms, data structure, and methodological techniques are explained in Sections 4 and 5. An example is reported in Section 6. Finally we present our conclusions and discuss further work. Proofs and further details are reported in [8].

2   Probabilistic Transition Systems

Probabilistic transition systems (PTS for short) generalise the well-known transition systems with probabilistic information. In a PTS, a transition does not lead to a single state but to a probability space whose sample space is a set of states. The model we define is widely used (see, e.g. [28,5,23].) and is also known as Markov decision processes [27]. We consider in addition a function that labels each state with a property assumed to be valid in this state.

Let $\text{Distr}(\Omega)$ be the set of all discrete probability distributions over the sample space Ω. Let PF be a set of propositional formulas closed under $\wedge$ and $\neg$.

Definition 1. *A probabilistic transition system (PTS for short) is a structure* $T = (S, \longrightarrow, f)$ *where S is a set of states,* $\longrightarrow \subseteq S \times \text{Distr}(S)$ *is the transition relation, and $f : S \to$ PF is a proposition assignment. We write $s \longrightarrow \pi$ if $(s, \pi) \in \longrightarrow$, and $s \longrightarrow$ if there is a π such that $s \longrightarrow \pi$; otherwise, we write $s \not\longrightarrow$ and call s a* sink state. *A PTS is said to be a* fully probabilistic transition system *(FPTS for short) if whenever $s \longrightarrow \pi$ and $s \longrightarrow \rho$ then $\pi = \rho$. It will be convenient to distinguish an initial state $s_0 \in S$. In this case we call the structure (T, s_0), a rooted (fully) probabilistic transition system. A proposition $g \in$ PF is satisfied in state s, notation $s \models g$, whenever $f(s) \Rightarrow g$ holds is a tautology.*

Example 1. Consider a system that either increments a counter with probability 0.5 or it deadlocks with probability 0.5 while the counter is smaller than 20. Formally, it can be modelled by a PTS Counter $= (S, \longrightarrow, f)$ where $S = \{a, b\} \times \{0 \ldots 20\}$, $f(s, i) = (s \wedge x = i)$, and

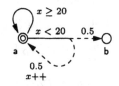

Fig. 1.

$$(a, 20) \longrightarrow \{(a, 20) \mapsto 1\}$$
$$(a, i) \longrightarrow \{(a, i + 1) \mapsto 0.5, (b, i) \mapsto 0.5\} \quad \text{if } i < 20$$

A symbolic representation of this PTS is depicted in Fig. 1.   □

Let $T = (S, \longrightarrow, f)$. A *simple path starting from* $s_0 \in S$ in T is a finite sequence of S-states, $\sigma = s_0 s_1 s_2 \ldots s_n$, where for each $0 \leq i < n$ there exists $\pi_i \in \text{Distr}(S)$ such that $s_i \longrightarrow \pi_i$ and $\pi_i(s_{i+1}) > 0$. Let $\sigma(i)$ denote the state in the i-th position. Let $|\sigma|$ be the length of σ. Let $first(\sigma) = \sigma(1)$ and $last(\sigma) = \sigma(|\sigma|)$. We let *s-paths*($T$) denote the sets of simple paths in T starting from any $s \in S$. A state t is *reachable* from other state s in T if there is $\sigma \in$ *s-paths*(T) with $s = first(\sigma)$ and $t = last(\sigma)$. Let *reach*(T, s) denote the set of all states reachable from s in T.

For any rooted FPTS (F, s), the *probability measure* $\mathsf{P}_{F,s}$ on the σ-algebra induced by (F, s) is the unique probability measure defined such that $\mathsf{P}_{F,s}(\sigma) =$ **if** ($s = s_0$) **then** $\pi_0(s_1) \cdot \pi_1(s_2) \cdot \ldots \cdot \pi_{n-1}(s_n)$ **else** 0. In particular, $\mathsf{P}_{F,s}(\sigma)$ is the probability of σ in F starting from s

Any given PTS T defines a set of *probabilistic executions*, each one obtained by iteratively scheduling one of the possible post-state distributions from each pre-state, starting from a given state $s_0 \in S$. Notice that the same state s of T may occur more than once during a probabilistic execution and each time a different distribution from s may be scheduled. In order to distinguish such occurrences every state s of a probabilistic execution is extended with the past history of s, that is, with the unique path leading from the start state to s.

Definition 2. *A probabilistic path of* T *is a FPTS* $F = (\text{s-paths}(T), \longrightarrow_F, f \circ last)$ *where* $q \longrightarrow_F \rho$ *implies* $last(q) \longrightarrow_T \pi$ *with* $\rho(qs) = \pi(s)$ *for all* $s \in S$. *If in addition, for all* $q \in$ *s-paths*(T) *such that* $|q| < i$, $last(q) \longrightarrow_T$ *implies that* $q \longrightarrow_F$, *then the rooted FPTS* (F, s_0) *is said to be a* probabilistic execution fragment *of length* i *of* T *starting from* $s_0 \in S$. *If* $i = \infty$ *then* (F, s_0) *is said to be a* probabilistic execution *of* T *starting from* $s_0 \in S$.

Denote by *paths*(T) the set of all probabilistic paths of T, by *execs*(T, s_0, i) the set of all probabilistic execution fragments of length i starting from s_0, and by *execs*(T, s_0) the set of all probabilistic executions of T starting from s_0.

Given a simple path $\sigma \in$ *s-paths*(T) define $\sigma^\uparrow \in$ *s-paths*(F) (F being a probabilistic path of T) such that $|\sigma^\uparrow| = |\sigma|$ and for all $0 < i \leq |\sigma|$, $\sigma^\uparrow(i) = \sigma(1) \ldots \sigma(i)$. We extend $^\uparrow$ to sets of simple paths in the usual way. Let $f \in$ PF and define $\Sigma_f \triangleq \{\sigma \in \text{s-paths}(T) \mid last(\sigma) \models f \text{ and } \forall 0 < i < |\sigma|. \sigma(i) \models \neg f\}$, i.e., Σ_f is the set of all minimal paths in T that end in final condition f. The *minimum* and *maximum probabilities* of reaching a final condition $f \in$ PF from an initial condition $i \in$ PF in a rooted PTS (T, s_0) are defined respectively by

$$\mathsf{P}^{\text{inf}}_{T,s_0}(i, f) \triangleq \inf \left\{ \mathsf{P}_{F,q}(\Sigma_f^\uparrow) \mid s \in \text{reach}(T, s_0), \; s \models i, \text{ and } (F, q) \in \text{execs}(T, s) \right\}$$

$$\mathsf{P}^{\text{sup}}_{T,s_0}(i, f) \triangleq \sup \left\{ \mathsf{P}_{F,q}(\Sigma_f^\uparrow) \mid s \in \text{reach}(T, s_0), \; s \models i, \text{ and } (F, q) \in \text{execs}(T, s) \right\}$$

Example 2. Consider the Counter of Example 1. Take the initial condition $i = (a \wedge x = 0)$ and the final condition $f = (b \wedge x \geq 15)$. The reader is invited to check that $\Sigma_f = \{(a, j)(a, j + 1) \ldots (a, i - 1)(a, i)(b, i) \mid 15 \leq i < 20 \wedge 0 \leq j \leq i\}$ and that $\mathsf{P}^{\text{inf}}_{T,(a,0)}(i, f) = \mathsf{P}^{\text{sup}}_{T,(a,0)}(i, f) = \frac{31}{2^{20}}$. □

3   Probabilistic Simulation

Probabilistic simulation [22,28] will be central to state the correctness of the technique proposed in this paper.

Definition 3. *Let* $C \subseteq S \times S$ *be a relation on states defining a* discrimination criterion. *R is a C-(probabilistic) simulation if, whenever sRt,*

1. $(s,t) \in C$, *and*
2. *if* $s \longrightarrow \pi$, *there exist* ρ *such that* $t \longrightarrow \rho$ *and* $\pi \sqsubseteq_R \rho$.

where $\pi \sqsubseteq_R \rho$ *if there is* $\delta \in \mathsf{Distr}(S \times S)$ *such that for all* $s,t \in S$, *(i)* $\pi(s) = \delta(s,S)$, *(ii)* $\rho(t) = \delta(S,t)$, *and (iii)* $\delta(s,t) > 0 \implies sRt$. *s is C-simulated by t, notation* $s \leq_C t$, *if there is a C-simulation R with sRt.*

Notice that whenever the discriminating criteria C is a preorder, so is $\leq_C$.

Our interest is to check when a PTS reaches a goal f starting from any state satisfying some initial condition i (i, f $\in$ PF). Consider the discriminating condition $(s,t) \in C_{i,f}$ defined by $(s \models f \implies t \models f)$ and $(s \models i \iff t \models i)$ We write only C whenever i and f are clear from the context. Simulations $\leq_C$, $\leq_{C^{-1}}$, and $\leq_{C \cap C^{-1}}$ are the relations needed to prove correctness of the technique.

We are interested in whether the probability of reaching a particular final condition f from any (reachable) state satisfying a given initial condition i is smaller or greater than a given value p. The next theorem states that if a PTS T_1 satisfies this property, and another T_2 $(C \cap C^{-1})$-simulates T_1, then T_2 also satisfies the property.

Theorem 1. *Let* (T_1, s_0^1) *and* (T_2, s_0^2) *be two rooted PTSs such that none of them contains a sink state. Then*

1. $(T_1, s_0^1) \leq_C (T_2, s_0^2)$ *implies* $P_{T_1,s_0^1}^{\mathrm{sup}}(i,f) \leq P_{T_2,s_0^2}^{\mathrm{sup}}(i,f)$.
2. $(T_1, s_0^1) \leq_{C^{-1}} (T_2, s_0^2)$ *implies* $P_{T_1,s_0^1}^{\mathrm{inf}}(i,f) \geq P_{T_2,s_0^2}^{\mathrm{inf}}(i,f)$.
3. $(T_1, s_0^1) \leq_{C \cap C^{-1}} (T_2, s_0^2)$ *implies* $P_{T_1,s_0^1}^{\mathrm{sup}}(i,f) \leq P_{T_2,s_0^2}^{\mathrm{sup}}(i,f)$ *and* $P_{T_1,s_0^1}^{\mathrm{inf}}(i,f) \geq P_{T_2,s_0^2}^{\mathrm{inf}}(i,f)$.

The requirement that every state has a transition is not really harmful as each sink state can always be completed with a self-looping transition without affecting the properties of the original PTS.

The proposed technique is based on successive refinements of a coarse abstraction of the original PTS T. Each refinement is an abstraction of the next (finer) refinement in which T is the finest one. In the following, we state that the refinement operation preserves simulation. A consequence of Theorem 1 is that if a given abstraction satisfy the desired reachability property, so does T.

Definition 4. *Let* $\mathcal{A} = (A_i)_I$ *be a partition of S, i.e., for all* $i,j \in I$, $A_i \cap A_j \neq \emptyset \iff i = j$, *and* $\bigcup_I A_i = S$. *Let* $T = (S, \longrightarrow, f)$. *The quotient PTS according to* $\mathcal{A}$ *is defined by* $T/\mathcal{A} = (\mathcal{A}, \longrightarrow_\mathcal{A}, f_\mathcal{A})$, *where*

1. $A \longrightarrow_{\mathcal{A}} (\pi/\mathcal{A})$ if $\exists s \in A.\ s \longrightarrow \pi$ and $\forall A' \in \mathcal{A}.\ (\pi/\mathcal{A})(A') \triangleq \sum_{s' \in A'} \pi(s')$,
2. $f_{\mathcal{A}}(A) \triangleq \bigwedge_{s \in A} f(s)$.

For a rooted PTS (T, s_0), its quotient is given by $(\mathsf{T}, s_0)/\mathcal{A} \triangleq (\mathsf{T}/\mathcal{A}, A)$ provided $s_0 \in A \in \mathcal{A}$.

We say that $\mathsf{T}/\mathcal{A}$ is a C-*abstraction* of T if for all $A \in \mathcal{A}$, and $s, t \in A$, $(s \models \mathsf{i} \iff t \models \mathsf{i})$. We say that $\mathsf{T}/\mathcal{A}$ is a $(\mathsf{C} \cap \mathsf{C}^{-1})$-*abstraction* of T if for all $A \in \mathcal{A}$, and $s, t \in A$, $(s \models \mathsf{f} \iff t \models \mathsf{f})$, and $(s \models \mathsf{i} \iff t \models \mathsf{i})$.

Theorem 2. *Let* (T, s_0) *be a rooted PTS. Let* $\mathcal{B}$ *be a refinement of* $\mathcal{A}$ *(i.e.,* $\mathcal{B}$ *is also a partition of* S *such that* $\forall B \in \mathcal{B}.\ \exists A \in \mathcal{A}.\ B \subseteq A$*).*

1. *if* $\mathsf{T}/\mathcal{A}$ *is a* C-*abstraction of* T *then (a)* $(\mathsf{T}, s_0)/\mathcal{B} \leq_{\mathsf{C}} (\mathsf{T}, s_0)/\mathcal{A}$, *and (b)* $\mathsf{T}/\mathcal{B}$ *is also a* C-*abstraction of* T.
2. *if* $\mathsf{T}/\mathcal{A}$ *is a* $(\mathsf{C} \cap \mathsf{C}^{-1})$-*abstraction of* T *then (a)* $(\mathsf{T}, s_0)/\mathcal{B} \leq_{\mathsf{C} \cap \mathsf{C}^{-1}} (\mathsf{T}, s_0)/\mathcal{A}$, *and (b)* $\mathsf{T}/\mathcal{B}$ *is also a* $(\mathsf{C} \cap \mathsf{C}^{-1})$-*abstraction of* T.

Example 3. Let

$$\mathcal{A} = \{\{(a,0)\}, \{(a,i) \mid 1 \leq i < 15\}, \{(a,i) \mid 15 \leq i < 20\}, \{(a,20)\},$$
$$\{(b,i) \mid 0 \leq i < 15\}, \{(b,i) \mid 15 \leq i < 20\}, \{(b,20)\}\}$$

Then $\mathsf{Counter}/\mathcal{A} = (\mathcal{A}, \longrightarrow_{\mathcal{A}}, f_{\mathcal{A}})$, with $\longrightarrow_{\mathcal{A}}$ defined by (see also Fig. 2)

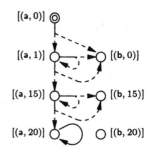

$$[(a,0)] \longrightarrow_{\mathcal{A}} \{[(a,1)] \mapsto 0.5, [(b,0)] \mapsto 0.5\}$$
$$[(a,1)] \longrightarrow_{\mathcal{A}} \{[(a,1)] \mapsto 0.5, [(b,0)] \mapsto 0.5\}$$
$$[(a,1)] \longrightarrow_{\mathcal{A}} \{[(a,15)] \mapsto 0.5, [(b,1)] \mapsto 0.5\}$$
$$[(a,15)] \longrightarrow_{\mathcal{A}} \{[(a,15)] \mapsto 0.5, [(b,15)] \mapsto 0.5\}$$
$$[(a,15)] \longrightarrow_{\mathcal{A}} \{[(a,20)] \mapsto 0.5, [(b,15)] \mapsto 0.5\}$$
$$[(a,20)] \longrightarrow_{\mathcal{A}} \{[(a,20)] \mapsto 1\}$$

Fig. 2.

where $[s]$ denotes the class of s, i.e., the set in $\mathcal{A}$ such that $s \in [s]$. Notice that $\mathsf{Counter}/\mathcal{A}$ is indeed a $(\mathsf{C} \cap \mathsf{C}^{-1})$-abstraction of $\mathsf{Counter}$ with i and f as before. In addition, $P^{\inf}_{\mathsf{T}, [(a,0)]}(\mathsf{i}, \mathsf{f}) = 0$ and $P^{\sup}_{\mathsf{T}, [(a,0)]}(\mathsf{i}, \mathsf{f}) = \frac{1}{4}$, which is a sound approximation of the actual solution (see Example 2).   □

4   Model-Checking and Partitioning

To perform model checking, the require the PTS under study to be finite. In order to describe the encoding of PTS we need some particular notation. If $s \longrightarrow \pi$, we call the pair (s, π) a *nail*. Nails have the same functionality as *probabilistic states* in alternating models; they are depicted with black boxes in

Fig. 3. Let $T = (S, \longrightarrow, f)$ be a PTS. From now on we assume the initial and final conditions i and f are atomic propositions and for all $s \in S$, $f(s)$ is either *true*, i, f or $i \wedge f$. T will also be described in an equivalent manner by a tuple (S, N, Org, τ, f) where N is the set of nails, $Org : N \to S$ associates to each nail (s, π) its *origin state* s, and $\tau : N \to \text{Distr}(S)$ associates its distribution π. Let $\mathcal{N}(s) = \{n \in N \mid Org(n) = s\}$ be the set of *outgoing nails* of state s. Notice that nails having different origin states can share the same distribution.

Computing $\mathsf{P}^{\text{inf}}(i)$ and $\mathsf{P}^{\text{sup}}(i)$. For the rest of this section we wil use the shorthand $\mathsf{P}^{\text{inf}}(s)$ and $\mathsf{P}^{\text{sup}}(s)$ for $\mathsf{P}^{\text{inf}}_{T,s}(s, f)$ and $\mathsf{P}^{\text{sup}}_{T,s}(s, f)$, respectively.

According to [5], the sets of states for which $\mathsf{P}^{\text{inf}}(s) = 0$ or $\mathsf{P}^{\text{sup}}(s) = 0$ can be computed by resorting to simple fixpoint computations on graphs, whereas the infimum and supremum probabilities of the other states satisfies the following equations:

$$\mathsf{P}^{\text{inf}}(s) = \min_{n \in \mathcal{N}(s)} \sum_{s' \in S} \tau(n)(s')\mathsf{P}^{\text{inf}}(s') \tag{1}$$

$$\mathsf{P}^{\text{sup}}(s) = \max_{n \in \mathcal{N}(s)} \sum_{s' \in S} \tau(n)(s')\mathsf{P}^{\text{sup}}(s')1 \tag{2}$$

To solve such a system, two methods have been explored: one can either transform such a system into a linear programming problem, and use classical techniques of linear programming, or consider the system as a fixpoint equation, and compute its least fixpoint by iterative methods. The solving method is however not the aim of this paper. We choosed linear programming with exact arithmetic, in order to avoid numerical problems and to get exact results.

Partitioning and complexity of the analysis. Basically, the two sources of complexity in these systems of equations are first the number of states of the PTS, which is of the same order as the number of variables, and then the number of nails, which gives the number of linear expressions in min or max expressions.

Partitioning the state space allows to address the first source of complexity. The question is how to proceed with the nails. Consider the PTS depicted in Fig. 3(a). The first effect of the abstraction is that several edges outgoing from a nail will lead to the same class; we have to merge these edges and add their probabilities, as shown on Fig. 3(b), where s_0, s_1 and s_2, s_3 are merged into equivalence classes k_0 and k_1. A second effect is that this operation makes some nails become equivalent, as (s_0, a_0) and (s_0, a_1) on Fig. 3(b). This effect is our main point: we expect that partitioning the state space will equate many nails and therefore address the second source of complexity. Such a situation is very likely to happen in systems that are specified in a symbolic way using data variables (see example 1).

5   Algorithms and Data Structures

Representation of states, transition relation, and partitions. As stated in the introduction, we use BDDs to represent sets of states, and ADDs to represent the transition relation.

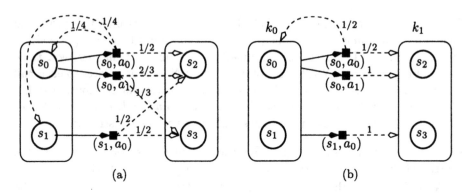

Fig. 3. A concrete PTS and its abstraction

As S is finite, we can encode each state $s \in S$ by a Boolean vector $\mathbf{s}$ of length $n = \lceil \log_2 |S| \rceil$. We then use BDDs to represent (sets of) states. Similarly, we use ADDs to represent the function τ, which belongs to the space $N \to (S \to [0,1])$, isomorphic to the space $N \times S \to [0,1]$. In order to encode nails, we use an auxiliary set A that solves in each state the nondeterministic choice on its outgoing nails. Let $p = \log(\max_{s \in S} |\mathcal{N}(s)|)$ and $\mathbb{B} = \{0,1\}$. We consider that $S \subseteq \mathbb{B}^n$, $A \subseteq \mathbb{B}^p$, $N \subseteq S \times A$, and $\tau : S \times A \times S \to [0,1]$. τ is then represented by an ADD using unprimed variables, auxiliary and primed variables, noted $\overrightarrow{\mathbf{s}}$, $\overrightarrow{\mathbf{a}}$, $\overrightarrow{\mathbf{s'}}$. The Boolean vectors $\mathbf{s}$ and $\mathbf{s'}$ represent states s and s', using respectively unprimed and primed variables. Each nail $n \in \mathcal{N}(s)$ outgoing from a state s is thus encoded by a pair $(\mathbf{s}, \mathbf{a}) \in S \times A$. Fig. 4(a) shows the ADD representing the system of Fig. 3(a).

A partition of a finite set S is defined by a set K of *classes*, and a function $def : K \to 2^S$ such that: $\bigcup_{k \in K} def(k) = S$ and $\forall k \neq k' : def(k) \cap def(k') = \emptyset$ (and $\forall k : def(k) \neq \emptyset$). As usual, sets are represented by BDDs. In order to use classes k in BDDs, we use Boolean vectors $\mathbf{k}, \mathbf{k'} \in \mathbb{B}^n$, represented with variables $\overrightarrow{\mathbf{k}}$ and $\overrightarrow{\mathbf{k'}}$.

Simplification of a PTS and Boolean analysis. Before performing any abstraction, we first try to simplify the PTS, using conditions i and f. Obviously, states that are not reachable from states satisfying i cannot influence the value of $\mathsf{P}^{\inf}(i)$ and $\mathsf{P}^{\sup}(i)$, so we can discard them from S and simplify τ. In a different spirit, the states s satisfying $\mathsf{P}^{\inf}(s) = \mathsf{P}^{\sup}(s) = 0$ or $\mathsf{P}^{\inf}(s) = \mathsf{P}^{\sup}(s) = 1$ can be respectively gathered in a *safe* partition or included in the final partition. These partitions are then transformed in a sink state by appropriately changing τ, since their outgoing transitions are irrelevant for the computation. Afterwards, a new reachability analysis allows to further simplify the state space and the transition function. The computation of such states can be done by fixpoint computations with BDDs, by considering a suitable Boolean abstraction of a PTS.

Notice that we do not necessarily reduce the number of nodes of BDDs and ADDs by the above simplifications. However, futile computations are avoided by restricting partitioning only to the relevant states. Boolean fixpoint computa-

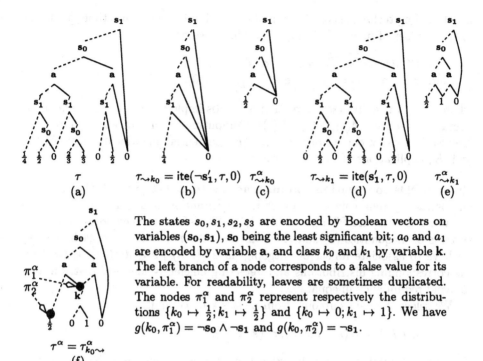

The states s_0, s_1, s_2, s_3 are encoded by Boolean vectors on variables (s_0, s_1), s_0 being the least significant bit; a_0 and a_1 are encoded by variable a, and class k_0 and k_1 by variable k. The left branch of a node corresponds to a false value for its variable. For readability, leaves are sometimes duplicated. The nodes π_1^α and π_2^α represent respectively the distributions $\{k_0 \mapsto \frac{1}{2}; k_1 \mapsto \frac{1}{2}\}$ and $\{k_0 \mapsto 0; k_1 \mapsto 1\}$. We have $g(k_0, \pi_1^\alpha) = \neg s_0 \wedge \neg s_1$ and $g(k_0, \pi_2^\alpha) = \neg s_1$.

Fig. 4. Representation and abstraction of the PTS of Fig. 3 with ADDs

tions are also used on PTS abstracted by partitioning in order to compute the set of classes k for which $P^{\mathrm{inf}}(k) = 0$ or $P^{\mathrm{sup}}(k) = 0$, which is required to solve equations (1) and (2) in Section 4.

Abstraction of a PTS by partitioning. The problem is the following: given a system $(S, N \subseteq S \times A, Org, \tau)$ and a partition (K, def) of S, compute an *abstract* system $(S^\alpha, N^\alpha, Org^\alpha, \tau^\alpha)$ with

$$S^\alpha = K \ , \ N^\alpha = N \ , \ \tau^\alpha : \quad N \quad \rightarrow \mathrm{Distr}(K)$$
$$(s, a) \mapsto \lambda k.(\textstyle\sum_{s' \in def(k)} \tau(s, a, s'))$$

We do not specify the function Org^α since it is not necessary to compute it.

To compute τ^α with ADDs, we first transform the summation indexed by $s' \in def(k)$ into an unconstrained summation. For $k' \in K$, define the ADD

$$\tau_{\leadsto k'}(s, a, s') = \mathrm{ite}(s' \in def(k'), \tau(s, a, s'), 0)$$

where ite is the if-then-else operator on ADDs (see Figs. 4(b) and (d)). Then, for every (s, a, k'), $\tau^\alpha(s, a, k') = \sum_{s'} \tau_{\leadsto k'}(s, a, s')$, which we note $\tau_{\leadsto k'}^\alpha(s, a)$. This unconstrained summation on all valuations taken by primed variables correspond exactly to the *existential quantification* of primed variables in the ADD $\tau_{\leadsto k'}$, as defined for instance in the library CUDD [30]. This operation benefits from the usual caching techniques of BDDs, and can be implemented in a time quadratic in the number of nodes of the input graph. So we have

$$\tau^\alpha_{\rightsquigarrow k'}(s,a) = \sum_{\vec{s}} \tau_{\rightsquigarrow k'}(s,a,s')$$
$$\text{and } \tau^\alpha(s,a,k') = \sum_{k' \in K} \text{ite}(k', \tau^\alpha_{\rightsquigarrow k'}(s,a), 0)$$

where $\sum_{k' \in K}$ is a disjoint summation on ADDs implemented by a cascade of ite operators (see Figs. 4(c), (e), and (f)). Computing τ^α in that way requires $|K|$ intersection operations (to obtain $\tau_{\rightsquigarrow k}$), $|K|$ existential quantifications on ADDs, and $|K|$ applications of the ite operator.

From ADDs to equations on abstract system. Let $(S^\alpha, N^\alpha, Org^\alpha, \tau^\alpha)$ be an abstract system defined by a partition with initial class k_i and final class k_f. We want to compute $\mathsf{P}^{\text{inf}}(k_i)$ and $\mathsf{P}^{\text{sup}}(k_i)$. Therefore we need to generate the systems of inequalities (1) and (2) from the ADD τ^α. That is, to each class k, we have to associate its set of outgoing nails $\{(s,a) \mid s \in def(k)\}$, extracting the corresponding distributions and detecting efficiently identical distributions to avoid redundancy in equations.

We select the nails outgoing from a class k by computing

$$\tau^\alpha_{k \rightsquigarrow}(s,a,k') = \text{ite}(s \in def(k), \tau^\alpha(s,a,k'), 0)$$

The important point now for the extraction of the distributions is that we require the variables $\vec{k'}$ to be ordered below the variables $\vec{s}$ and $\vec{a}$ in ADDs. This allows to extract the distributions by performing a depth-first search of the graph rooted at $\tau^\alpha_{k \rightsquigarrow}$, stopping as soon as a node indexed by a variable belonging to $\vec{k'}$ is encountered. Such a node corresponds to a distribution (Fig. 4(f)). Because of this variable ordering and the sharing of nodes the ADD $\tau^\alpha_{k \rightsquigarrow}$, the set of its *different* distributions can be obtained for free by a simple graph algorithm.

The third step, generating a linear expression from an ADD representing a distribution, is done by enumerating the valuations on variables $\vec{k'}$ that leads to a non-zero leaf, c.f. Fig. 4(f) and the explanations. We resort then to section 4 to solve the system of equations.

Automatic partition refinement. The choice of a suitable abstraction is a difficult problem, because only the results of the analysis can decide whether the abstraction offers enough precision to check the intended property. This is why we have chosen an incremental partitioning method.

The verification starts with a rough partition of the system. If the analysis of this abstract PTS allows to conclude that the property is satisfied by the concrete PTS, the verification process is finished. Otherwise, a partition refinement step is performed in order to obtain more precise information. This process is iterated up to success or until all classes of the partition are stable. If this last situation occurs, we can conclude that the property is false and extract a counter-example path.

The initial partition contains three distinguished classes: the safe, initial, and final classes, denoted k_s, k_i, and k_f, with $def(k_s) = \{s \mid \mathsf{P}^{\text{sup}}(s) = 0\}$, $def(k_i) = i$, and $def(k_f) = f$. The safe and final classes are never split. As our tool allows to specify processes that combines an explicit control structure and operations on data variables, we use this control structure to partition the remaining state space.

Our refinement method tries to stabilize classes, by separating concrete states in a class that have different future, as do all partition refinement methods based on a bisimulation criteria [6,1,31], and most of those dealing with infinite state systems [29,21]. We implement this idea by considering the set of states $g(k, \pi^\alpha) \subseteq def(k)$ in class k that can lead to the abstract distribution π^α. $g(k, \pi^\alpha)$ can be seen as the *guard* of the distribution π^α in class k. For instance, in Fig. 3(b), if π_1^α denotes the distribution attached to the nail (s_0, a_0), then $g(k_0, \pi_1^\alpha) = \{s_0\}$, and if π_2^α denotes the distribution associated to nails (s_0, a_1) and (s_1, a_0), $g(k_0, \pi_2^\alpha) = \{s_0, s_1\}$. If such a guard is neither empty, nor equal to the definition of the class k, then the class k can be safely split into two classes k' and k'' according to this *discriminating* guard, with $def(k') = g(k, \pi^\alpha)$ and $def(k'') = def(k) \setminus g(k, \pi^\alpha)$, because states in class k' are certainly not bisimilar to states in class k''. A guard $g(k, \pi^\alpha)$ is obtained by computing the union of paths in the ADD $\tau_{k \leadsto}^\alpha$ that leads from the root node to the node representing the distribution π^α (Fig. 4(f)). Such an operation can again be implemented with a complexity linear in the number of nodes of the ADD $\tau_{k \leadsto}^\alpha$.

Our global strategy for refinement tries, between each analysis step, to split once every class for which there exist a guard. After a partition refinement, a new abstract transition function τ^α has to be computed. When a class k has not been split, the ADDs $\tau_{\leadsto k}$ and $\tau_{\leadsto k}^\alpha$ are reused; otherwise, we need to recompute them, as well as the ADDs τ^α and $\tau_{k \leadsto}^\alpha$. So the refinement process require $\mathcal{O}(|K|)$ BDDs operations.

Conclusion. The algorithms presented in this section allows to partition and to refine an abstract PTS with $\mathcal{O}(|K|)$ BDDs operations; the complexity of these operations is in turn linear or quadratic in the number of *nodes* of the input diagrams.

6   Example

The Bounded Retransmission Protocol (BRP) [15,12,7] has become a nice benchmark example as it is simple to understand, yet its overall behaviour is not trivial. The BRP is based on the alternating bit protocol but allows for a bounded number of retransmissions of a *chunk*, i.e., part of a file, only. So, eventual delivery is not guaranteed and the protocol may abort the file transfer. By using our technique, we are able to quantify the probability of such abortion.

The protocol consists of a Sender and a Receiver exchanging data via two unreliable (lossy) channels, K and L. The Sender reads a file to be transmitted (which is assumed to be divided in N chunks) and sets the retry counter to 0. Then it sends the elements of the file one by one over K to the Receiver. A *frame* sent through channel K consists of three bits and a chunk. The first bit indicates

Sender:

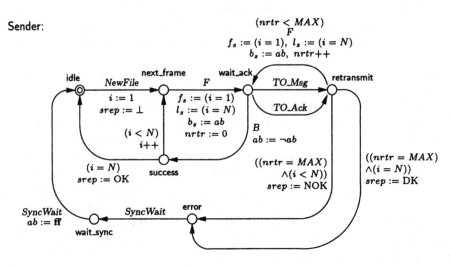

Receiver:

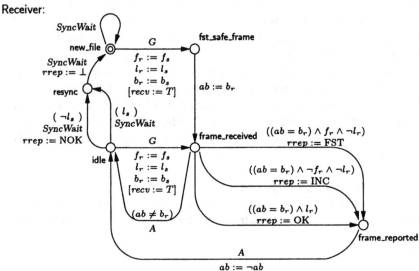

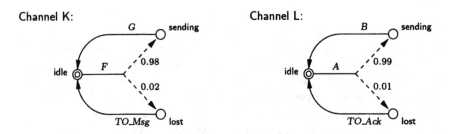

Fig. 5. PTS model of the bounded retransmission protocol

whether the chunk is the first element of the file, the second one indicates if it is the last one, and the third bit, the so-called alternating bit, is used to guarantee that data is not duplicated. After sending the frame, the Sender waits for an acknowledgement or for a timeout. In case of acknowledgement, if it corresponds to the last chunk, the sending client is informed of correct transmission (signal OK); otherwise the next element of the file is sent. If a timeout occurs, the frame is resent (after the counter for the number of retries is incremented), or the transmission of the file is broken off. The latter occurs if the retry counter exceeds its maximum value MAX. In this case the sender client is informed whether the Sender did not complete the transmission (NOK), or whether it sent the last chunk but it was never acknowledge (DK) in which case the success of the transmission is unknown. Afterwards, and before sending a new file, the Sender waits enough time to ensure that the Receiver has properly reacted to the communication break.

The Receiver waits for a frame to arrive. This frame is delivered at the receiving client informing whether it is the first (FST), an intermediate (INC), or the last one (OK). Afterwards, an acknowledgement is sent over L to the Sender. Then the Receiver simply waits for more frames to arrive. The receiver remembers whether the previous frame was the last element of the file and the expected value of the alternating bit. Each frame is acknowledged, but it is handed over to the receiving client only if the alternating bit indicates that it is new. Note that (only) if the previous frame was last of the file, then a fresh frame will be the first of the subsequent file and a repeated frame will still be the last of the old file. If a long enough time had passed since the last frame was received, the Receiver assumes that the normal communication flow broke down. If this happen, the receiving client is informed, provided the last element of the file has not yet been delivered. Since our model does not consider time, we assume that premature timeouts are not possible and that the Sender and Receiver always re-synchronise properly after normal communication is broken.

The description of the components of the protocol in terms of PTS is given in Fig. 5. It abstracts from the data that is being transmitted. The components synchronise through common alphabet (a la CSP [19]). Notice that the only probabilistic features are those occurring in the medium. In this model we assume that a frame is lost with probability 0.02, and acknowledgement is lost with probability 0.01.

A checking automaton (Fig. 6) ensures that the transmitted file is invariant for the property under study, i.e, the property is only interesting for exactly one file transmission. Notice that the checking automata selects an arbitrary file to test. We study several properties. The considered initial condition is test@Check ∧ next_frame@Sender ∧ (i@Sender = 1). The different final conditions are listed in Table 1. Notice the flag *recv* at the Receiver side; it is used to register that the last sent file has actually started to be received. Properties A and B define the minimal correctness requirement of the protocol. They should *not* be valid.

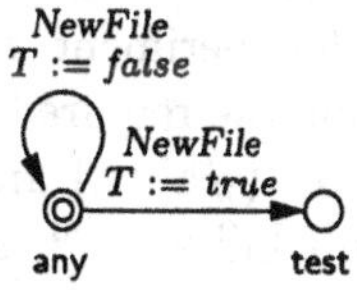

Fig. 6.

Table 1. Reachability Conditions

	Final condition	Meaning of the property
A	$(srep@\text{Sender} = \text{NOK}) \land$ $(rrep@\text{Receiver} = \text{OK}) \land$ $(recv@\text{Receiver})$	The Sender reports a certain unsuccessful transmission but the Receiver got the complete file. (The probability should be 0!)
B	$(srep@\text{Sender} = \text{OK}) \land$ $\neg(rrep@\text{Receiver} = \text{OK}) \land$ $(recv@\text{Receiver})$	The Sender reports a certain successful transmission but the Receiver did not get the complete file. (The probability should be 0!)
1	error@Sender	The Sender does not report a successful transmission
2	error@Sender $\land$ $(srep@\text{Sender} = \text{DK})$	The Sender reports an uncertainty on the success of the transmission
3	error@Sender $\land$ $(srep@\text{Sender} = \text{NOK}) \land$ $(i@\text{Sender} > 8)$	The Sender reports a certain unsuccessful transmission after transmitting half a file
4	$\neg(srep@\text{Sender} = \bot) \land$ $\neg(recv@\text{Receiver})$	The Receiver does not receive any chunk of a file, i.e., the first message never arrives. "$\neg(srep@\text{Sender} = \bot)$" ensures that the Sender did try to send a chunk

Properties 1 to 3 are concerned with transmissions that the Sender does not consider successful while property 4 considers an attempt for transmission with no reaction at the Receiver side.

The exercise we perform is to try to find the minimum number of retransmissions (MAX) that satisfies our probabilistic requirements for these properties when the transmitted file has length $N = 16$. Table 2 reports these results. Some remarks are in order. Each row in Table 2 reports a different instance according to the maximum number of retransmission MAX which is specified in the first column. The second column reports the number of reachable states in the respective instance (#reach.), and the third one the number of relevant states, i.e., reachable states that may lead to a state satisfying the final condition (#relev).

For each property we tried two different values of desired probability. Thus, for instance, property 1 is require to hold with probability less than 0.05 in the first experiment and than 0.01 in the second one. The least three columns report the last possible refinement together with its respective convergence value. For each experiment we report the number of refinements (#refin.) necessary to conclude the required property and for this last refinement, the number of abstract states (#abst.) and the upper bound for the actual minimum and maximum probability (P^{inf} and P^{sup}, respectively). We also report whether the property holds ($\sqrt{}$) or not ($\times$) on the verdict columns (Verd.)

Notice that, in this example, the proposed method do the actual verdict on an abstract state space which is, on average, around 20 times smaller than the concrete reachable state space. In particular it performed quite well for Properties 2 and 3 in the larger systems ($MAX \geq 3$). We have experience two

Table 2. Results in a BRP with file length = 16

MAX	#reach.	#relev.	#refin.	#abat.	pinf	psup	Verd.	#refin.	#abat.	pinf	psup	Verd.	#refin.	#abat.	pinf = psup
Property 1:					prob. ≤ 10⁻³					prob. ≤ 10⁻⁵				Convergence	
0	1174	269	1	12	0.0298		1 ×	1	12	0.0298		1 ×	88	99	0.383717
1	2068	697	6	35	1.5679e-03		1 ×	1	17	6.01899e-04		1 ×	92	247	0.0141144
2	2962	1125	92	404	4.22546e-04	4.24145e-04	√	3	27	1.20404e-05		1 ×	94	411	4.23333e-04
3	3856	1553	94	564	1.25943e-05	1.2642e-05	√	76	459	1.02284e-05		1 ×	96	575	1.26178e-05
4	4750	1981	95	724	3.75311e-07	3.76733e-07	√	95	724	3.75311e-07	3.76733e-07	√	97	739	3.76012e-07
5	5644	2409	95	884	1.11843e-08	1.12267e-08	√	95	884	1.11843e-08	1.12267e-08	√	97	903	1.12051e-08
Property 2:					prob. ≤ 10⁻⁴					prob. ≤ 10⁻⁶				Convergence	
0	1174	1134	85	99	0.0189293	0.0189293	×	85	99	0.0189293	0.0189293	×	85	99	0.0189293
1	2068	2028	87	244	8.76261e-04	8.76307e-04	×	87	244	8.76261e-04	8.76307e-04	×	89	247	8.76284e-04
2	2962	2922	8	54	0	2.64633e-05	√	89	404	2.64531e-05	2.64531e-05	×	91	411	2.64531e-05
3	3856	3816	9	72	0	8.88033e-06	√	10	76	0	7.88615e-07	√	93	575	7.88606e-07
4	4750	4710	9	81	0	2.64636e-05	√	11	93	0	2.64636e-07	√	96	739	2.35007e-08
5	5644	5604	9	82	0	2.64636e-05	√	11	97	0	7.88615e-07	√	99	903	7.00322e-10
Property 3:					prob. ≤ 10⁻³					prob. ≤ 10⁻⁵				Convergence	
0	1174	950	43	93	0.149825	0.149825	×	43	93	0.149825	0.149825	×	43	93	0.149825
1	2068	1844	45	229	6.15567e-03	6.15599e-03	×	45	229	6.15567e-03	6.15599e-03	×	47	232	6.15584e-03
2	2962	2738	42	360	0	1.85196e-04	√	47	379	1.85191e-04	1.85191e-04	×	49	386	1.85191e-04
3	3856	3632	59	519	0	5.52026e-06	√	59	519	0	5.52026e-06	√	68	540	5.52026e-06
4	4750	4526	43	371	0	3.04092e-04	√	46	379	0	1.64505e-06	√	102	693	1.64505e-07
5	5644	5420	52	455	0	8.8846e-06	√	52	455	0	8.8846e-06	√	128	848	4.90225e-09
Property 4:					prob. ≤ 10⁻³					prob. ≤ 10⁻⁵				Convergence	
0	1174	256	1	4	0.02	0.02	×	1	4	0.02	0.02	×	1	4	0.02
1	2068	465	1	6	4e-04	4e-04	√	1	6	4e-04	4e-04	×	1	6	4e-04
2	2962	674	1	6	0	4e-04	√	3	8	8e-06	8e-06	√	3	8	8e-06
3	3856	883	1	6	0	4e-04	√	3	8	0	8e-06	√	5	10	1.6e-07
4	4750	1092	1	6	0	4e-04	√	3	8	0	8e-06	√	7	12	3.2e-09
5	5644	1301	1	6	0	4e-04	√	3	8	0	8e-06	√	9	14	6.4e-11

Table 3. Performance (with time format "h:mm:ss.d")

MAX	Property 1 (≤ 10⁻⁵)	Converg.	Property 2 (≤ 10⁻⁶)	Converg.	Property 3 (≤ 10⁻⁵)	Converg.	Property 4 (≤ 10⁻⁵)	Converg.
0	0.5	11.6	13.3	13.5	8.7	8.8	0.4	0.4
1	0.6	2:25.6	3:54.1	4:11.9	1:49.4	2:02.2	0.5	0.5
2	1.2	8:27.8	11:02.8	11:24.8	5:16.9	5:55.2	0.6	0.6
3	8:50.5	17:05.8	6.3	22:44.9	11:29.2	15:30.7	0.7	0.7
4	27:19.9	28:52.1	8.8	35:50.7	5:58.0	40:52.9	0.7	0.8
5	41:09.3	45:05.0	10.2	52:21.0	11:03.0	1:31:14.7	0.8	0.9

different situations: either there is a gradual convergence to the infimum, but almost none to the supremum until an abrupt convergence in the last refinements (e.g. Property 1), or vice-versa (e.g Props. 2 and 3). The first case case may allow for an early rejection of the required property but would require many refinements if the property does hold (compare the number of refinements and abstract states of the × cases against the √ cases in Property 1). Instead, the second case will give an early report if the property holds (compare now the different results in Property 2).

The exercise of convergence is more costly as no criterion to stop the execution is provided and it proceeds until no more refinement is possible or the probability has definitely converged. At this maximum point notice that the state compression ratio is 10 times on average.

7   Concluding Remarks

In this article we introduced an efficient technique for quantitative model checking. The method relies on automatic abstraction of the original system. This allows to significantly reduce the size of the problem to which numerical analysis is applied in order to compute the quantitative factor of the property under study. Since the numerical analysis is the most costly part of the whole process this reduction is of high importance. This reduction is achieved first because bisimilar states are never distinguished, and secondly because using incremental abstraction refinement and confronting the analysis against a desired (or undesired) probability allows prompt answers on very compact spaces.

The execution time is currently not the best as the tool should be optimised. Table 3 reports the tool performance for a set of properties[1]. The current implementation performs numerical analysis using linear programming techniques under exact rational arithmetics. This method is very fast (compared to the painfully slow iterative methods) and it does not suffer of numerical unstability since numbers are represented in its exact form. Two remarks are in order. First, numerical analysis is applied in each refinement step, which is inefficient since a refinement step may add only few partitions with low chances of sensibly affecting the result of the previous iteration. Second, the already mentioned asymmetric convergence in which only the minimum or the maximum gradually converges to the actual value while the other does not until the last refinements.

It is in our near future plans to develop efficiency improvements. One of these improvements concerns the refinement strategy and the suitable alternation of refinement and analysis that should be used. Another improvement would be to take advantage of the fact that probabilities usually appears only in some part of the modelled system: failures do not appear everywhere!

On a long term agenda, we plan to use this incremental refinement technique to check probabilistic timed automata. Model checking of PTCTL properties on such model was proven decidable by resorting to their region graphs [25]. However, region graphs are known to be impractical. Our technique would allow to generate progressively a minimal *probabilistic* model, in the spirit of [1].

Acknowledgements. We thank Holger Hermanns and Joost-Pieter Katoen for fruitful discussions.

References

1. R. Alur, C. Courcoubetis, N. Halbwachs, D. Dill, and H. Wong-Toi. Minimization of timed transition systems. In R. Cleaveland, ed., *Procs. of CONCUR 92*, Stony Brook, NY, LNCS 630, pp. 340–354. Springer, 1992.
2. A. Aziz, V. Singhal, F. Balarin, R.K. Bryton, and A.L. Sangiovanni-Vincentelli. It usually works:the temporal logics of stochastic systems. In P. Wolper, ed., *Procs. of the 7th CAV*, Liège, LNCS 939, pp. 155–165. Springer, 1995.

---

[1] Time measured on a Sun Enterprise E3500 under Solaris 5.6.

3. R.I. Bahar, E.A. Frohm, C.M. Gaona, G.D. Hachtel, E. Macii, A. Pardo, and F. Somenzi. Algebraic decision diagrams and their applications. *Formal Methods in System Design*, 10(2/3):171–206, 1997.

4. C. Baier, J.-P. Katoen, and H. Hermanns. Approximate symbolic model checking of continuous-time Markov chains. In J.C.M. Baeten and S. Mauw, eds., *Procs. of CONCUR 99*, Eindhoven, LNCS 1664, pp. 146–161. Springer, 1999.

5. A. Bianco and L. de Alfaro. Model checking of probabilistic and nondeterministic systems. In *Procs. 15th FSTTCS* , Pune, LNCS 1026, pp. 499–513. Springer, 1995.

6. A. Bouajjani, J. C. Fernandez, N. Halbwachs, P. Raymond, and C. Ratel. Minimal state graph generation. *Science of Computer Programming*, 18:247–269, 1992.

7. P.R. D'Argenio, J.-P. Katoen, T.C. Ruys, and J. Tretmans. The bounded retransmission protocol must be on time! In E. Brinksma, ed., *Procs. of the 3rd TACAS*, Enschede, LNCS 1217, pp. 416–431. Springer, 1997.

8. P.R. D'Argenio, B. Jeannet, H.E. Jensen, and K.G. Larsen. Reachability Analysis of Probabilistic Systems by Successive Refinements. CTIT Technical Report, 2001. To appear.

9. L. de Alfaro, M. Kwiatkowska, G. Norman, D. Parker, and R. Segala. Symbolic model checking of concurrent probabilistic processes using MTBDDs and the Kronecker representation. In Graf and Schwartzbach [11].

10. M. Fujita, P.C. McGeer, and J.C.-Y. Yang. Multi-terminal binary decision diagrams: An efficient data structure for matrix representation. *Formal Methods in System Design*, 10(2/3):149–169, April 1997.

11. S. Graf and M. Schwartzbach, eds. *Procs. of the 6th Workshop TACAS*, Berlin, LNCS 1785. Springer, 2000.

12. J.F. Groote and J. van de Pol. A bounded retransmission protocol for large data packets –A case study in computer checked algebraic verification–. In M. Wirsing and M. Nivat, eds., *Procs. of the 5th AMAST Conference*, Munich, LNCS 1101. Springer, 1996.

13. H.A. Hansson and B. Jonsson. A logic for reasoning about time and reliability. *Formal Aspects of Computing*, 6:512–535, 1994.

14. V. Hartonas-Garmhausen and S. Campos. ProbVerus: Probabilistic symbolic model mhecking. In In Katoen [24], pp. 96–110.

15. L. Helmink, M.P.A. Sellink, and F.W. Vaandrager. Proof-checking a data link protocol. In H. Barendregt and T. Nipkow, eds., *Procs. International Workshop TYPES'93*, Nijmegen, LNCS 806, pp. 127–165. Springer, 1994.

16. H. Hermanns. Personal communication, 2001.

17. H. Hermanns, J.-P. Katoen, J. Meyer-Kayser, and M. Siegle. A Markov chain model checker. In Graf and Schwartzbach [11], pp. 347–362.

18. H. Hermanns, J. Meyer-Kayser, and M. Siegle. Multi terminal binary decision diagrams to represent and analyse continuous time Markov chains. In B. Plateau, W.J. Stewart, and M. Silva, eds., *3rd Int. Workshop on the Numerical Solution of Markov Chains*, pp. 188–207. Prensas Universitarias de Zaragoza, 1999.

19. C.A.R. Hoare. *Communicating Sequential Processes*. Prentice-Hall International, Englewood Cliffs, 1985.

20. M. Huth and M. Kwiatkowska. Quantitative analysis and model checking. In *Procs. 12th Annual Symposium on Logic in Computer Science*, Warsaw. IEEE Press, 1997.

21. B. Jeannet. Dynamic partitioning in linear relation analysis. Application to the verification of reactive systems. *Formal Methods in System Design*, 2001. To appear.

22. B. Jonsson and K.G. Larsen. Specification and refinement of probailistic processes. In *Procs. 6th Annual Symposium on Logic in Computer Science*, Amsterdam, pp. 266–277. IEEE Press, 1991.

23. B. Jonsson, K.G. Larsen, and W. Yi. Probabilistic extensions in process algebras. In J.A. Bergstra, A. Ponse, and S. Smolka, eds., *Handbook of Process Algebras*, pp. 685–710. Elsevier, 2001.

24. J.-P. Katoen, ed. *Procs of the 5th ARTS*, Bamberg, LNCS 1601. Springer, 1999.

25. M. Kwiatkowska, G. Norman, R. Segala, and J. Sproston. Automatic verification of real-time systems with probability distributions. In Katoen [24], pp. 75–95.

26. K.G. Larsen and A. Skou. Bisimulation through probabilistic testing. *Information and Computation*, 94:1–28, 1991.

27. M.L. Puterman. *Markov Decision Processes: Discrete Stochastic Dynamic Programming*. John Wiley & Sons, 1994.

28. R. Segala. *Modeling and Verification of Randomized Distributed Real-Time Systems*. PhD thesis, Massachusetts Institute of Technology, 1995.

29. H. Sipma, T.E. Uribe, and Z. Manna. Deductive model checking. In R. Alur and T.A. Henzinger, eds. *Procs. of the 8th CAV*, New Brunswick, New Jersey, LNCS 1102. Springer, 1996.

30. F. Somenzi. CUDD: Colorado University Decision Diagram Package. ftp://vlsi.colorado.edu/pub.

31. R. F. Lutje Spelberg, W. J. Toetenel, and M. Ammerlaan. Partition refinement in real-time model checking. In A.P. Ravn and H. Rischel, eds., *Procs. of the 5th FTRTFT*, Lyngby, LNCS 1486, pp. 143–157. Springer, 1998.

Beyond Memoryless Distributions: Model Checking Semi-Markov Chains

Gabriel G. Infante López, Holger Hermanns, and Joost-Pieter Katoen

Formal Methods and Tools Group, Faculty of Computer Science
University of Twente, P.O. Box 217, 7500 AE Enschede, The Netherlands

Abstract. Recent investigations have shown that the automated verification of continuous-time Markov chains (CTMCs) against CSL (Continuous Stochastic Logic) can be performed in a rather efficient manner. The state holding time distributions in CTMCs are restricted to negative exponential distributions. This paper investigates model checking of semi-Markov chains (SMCs), a model in which state holding times are governed by general distributions. We report on the semantical issues of adopting CSL for specifying properties of SMCs and present model checking algorithms for this logic.

1 Introduction

Model checking is a technique that is more and more used to ascertain properties of computer software, hardware circuits, communication protocols, and so forth. In this approach, properties are specified via an appropriate temporal logic, such as CTL or LTL, while systems are represented as (usually finite) state-transition diagrams. More recently, model checking techniques have been extended to stochastic processes such as continuous-time Markov chains (CTMCs, for short). In particular, efficient verification algorithms have been developed for CSL (Continuous Stochastic Logic [3,4,5]), a stochastic variant of CTL. CSL supports the specification of sophisticated steady-state and time-dependent properties. CTMCs are widely used in practice, mainly because they combine a reasonable modelling flexibility with well-established efficient analysis techniques for transient and steady-state probabilities that form the basis for determining performance measures such as throughput, utilisation and latencies. The stochastic processes described by CTMCs are characterised by the fact that the state holding times, indicating the amount of time the system stays in a state, are restricted to negative exponential distributions. As a result of their so-called memoryless property, the probability of moving from one state to another is independent of the amount of time the system has spent in the current state.

Although exponential distributions appropriately model a significant number of phenomena – related to mass effects – of random nature, in many occasions they are inadequate to faithfully model the stochastic behaviour of the system under consideration. For example, file sizes of documents transferred via the Internet, cycle times in hardware circuits, timeouts in communication protocols,

L. de Alfaro and S. Gilmore (Eds.): PAPM-PROBMIV 2001, LNCS 2165, pp. 57–70, 2001.

human behaviour, hardware failures, and jitter in multi-media communication systems cannot be appropriately modelled. In order to model these phenomena in an adequate manner, *general* distributions such as heavy-tail [10] (for file sizes), deterministic (for cycle times and timeouts), log-normal (for human response behaviour [21]), Weibull (for hardware failures [20]), and normal distributions (for jitter [13]) are used. To adopt the model checking approach to these distributions, the simplest solution is to approximate general distributions by the mean times to absorption of a CTMCs with an absorbing state, representing a so-called phase-type distribution. Although the resulting CTMC can be analysed using the existing verification algorithms and prototype tools for CSL, such approximations (i) easily give rise to a state-space explosion – the number of states increases significantly with the accuracy of the approximation and the degree of determinism of the desired distribution – are (ii) not easy to handle in case of a choice between stochastic delays – a race condition between the entire approximated distributions is decisive – and (iii) require the ability to fit the desired distribution by an appropriate phase-type distribution – a non-trivial problem in general, see e.g., [2].

Therefore as an alternative approach, this paper investigates direct model checking of semi-Markov chains (SMCs, for short) [8,18], a natural extension of CTMCs in which state holding times are determined by *general* continuous distributions. First, the semantics of CSL on SMCs is studied. In particular, the formal characterisation of the CSL steady-state operator is adapted as limit state probabilities are not guaranteed to exist for finite-state SMCs, in contrast to finite-state CTMCs. Instead, the behaviour of SMCs on the long run is characterised using the average fraction of time the system resides in a state. For instance, the formula $S_{<0.01}(error)$ is valid in state s iff on the long run for at most 1% of the time on average the system is in an error state when starting in state s. For finite CTMCs this interpretation is equal to the characterisation using the limit state probabilities. Secondly, model checking algorithms are proposed to verify CSL over finite-state SMCs. Although long-run properties are semantically characterised in a slightly different way, they can be checked as for CTMCs: a graph analysis to determine the bottom strongly connected components and solving a linear system of equations for each such component suffice. (In the literature, strongly connected SMCs are also known as irreducible SMCs.) Time-bounded until formulas can be checked, like for CTMCs, by a reduction to transient analysis of SMCs. These include probabilistic timed reachability properties such as: can the system reach a goal-state within a certain time-bound with some minimal (or maximal) probability? Whereas such transient analysis for CTMCs can be solved via stable and efficient numerical techniques such as uniformisation, for SMCs it requires solving a set of non-trivial Volterra equations whose solution algorithms have a worst case time complexity of $\mathcal{O}(N^4)$, where N is the number of states of the SMC under consideration.

In the context of logical specification formalisms and automated verification, stochastic processes with general distributions have received scant attention in the literature so far. Three related works are known to the authors. Van Hung and Chaochen [17] have defined a probabilistic variant of the duration calculus to express properties over SMCs, but did not report on any verification algorithms. De Alfaro [12] discusses model checking of long-run average properties and expected reachability times on semi-Markov decision processes. These models can be considered as SMCs extended with non-determinism. Time-bounded formulas are not considered. Kwiatkowska *et al.* [19] have recently considered the verification of a stochastic variant of timed automata, with clocks that are governed by general distributions, against properties in probabilistic timed CTL. They show that a finite-state semantics of such timed automata can be obtained using the region-based technique [1] where regions are partitioned to cater for the stochastic behaviour. Due to the intrinsic complexity of the model checking algorithm, it seems practically infeasible.

Organisation of the paper. Section 2 introduces the basic concepts of SMCs. Section 3 recalls the logic CSL and defines the semantics of CSL over SMCs. Model checking algorithms for long-run properties and time-bounded until formulas are described in Section 4. Section 5 concludes the paper.

2   Semi-Markov Chains

A semi-Markov chain (SMC) can be considered as a Kripke structure in which the transitions are labelled by information about the speed at which the chain evolves from one state to another. In a SMC, the delay between two successive state changes can be *generally* distributed. This property has to be contrasted with continuous-time Markov chains (CTMCs) where these delays need to be governed by negative exponential distributions. In this section, we introduce the basic concepts of SMCs. A more thorough treatment of SMCs can be found in [8,18].

Semi-Markov chains. Let AP be a fixed, finite set of atomic propositions. A (labelled) SMC $\mathcal{M}$ is a tuple $(S, \mathbf{P}, \mathbf{Q}, L)$ where S is a finite set of states, $\mathbf{P} : S \times S \to [0, 1]$ is the *transition probability matrix* (satisfying $\sum_{s' \in S} \mathbf{P}(s, s') = 1$ for each s), $\mathbf{Q} : S \times S \times (\mathbb{R}_{\geq 0} \to [0, 1])$ is a matrix of continuous probability distribution functions (such that $\mathbf{P}(s, s') = 0$ implies $\mathbf{Q}(s, s', t) = 1$), and $L : S \to 2^{AP}$ is the labelling function. Function L assigns to each state $s \in S$ the set $L(s)$ of atomic propositions $a \in AP$ that are valid in s.

The intuitive interpretation of a SMC is as follows. There exists a transition from state s to s' (which possibly equals s) if and only if $\mathbf{P}(s, s') > 0$. Matrix $\mathbf{P}$ determines the (discrete) probabilistic behaviour when changing from one state to another, i.e., $\mathbf{P}(s, s')$ is the probability to move from state s to state s'. Note that this is identical to the probabilistic branching of a discrete-time Markov chain (DTMC); $(S, \mathbf{P}, L)$ is often called the *embedded* DTMC of SMC $\mathcal{M}$. Once a next state s' of state s has been selected, the state holding time of state s is determined according to the probability distribution function $\mathbf{Q}(s, s', t)$. Thus, $\mathbf{Q}(s, s', t)$ denotes the probability to move from state s to s' within at most

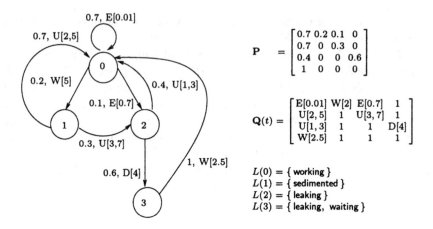

Fig. 1. A SMC describing a boiler.

t time-units, given that a transition from s to s' will be taken. A state s is absorbing if $\mathbf{P}(s, s) = 1$ and $\mathbf{Q}(s, s)$ is some arbitrary nontrivial distribution. The distribution function $\mathbf{H}$ of state s, defined by

$$\mathbf{H}(s,t) = \sum_{s' \in S} \mathbf{P}(s, s') \cdot \mathbf{Q}(s, s', t)$$

denotes the total holding time distribution in s regardless of which successor is selected.

We assume that the system will stay in a state with at least some non-zero probability, or more formally we demand for arbitrary s that there is some $t' > 0$ and some $\varepsilon > 0$ such that $\mathbf{H}(s, t') < 1 - \varepsilon$. We further require the mean of each state holding time distribution to be finite, i.e., $E[\mathbf{H}(s)] \neq \infty$.

Example 1. As a simple example of a SMC we model a boiler. The system can be in four different states, state 0 where the boiler is working properly, state 1, where the boiler has too much sediment that needs to be removed, state 2 where a pipe is leaking that either needs to be fixed or needs to be replaced, and finally state 3 where the system is waiting for a new pipe to arrive for replacement. The model is schematically depicted in Fig. 1, together with the matrices $\mathbf{P}$ and $\mathbf{Q}$ ('E' denotes an exponential distribution, 'U' a uniform distribution, 'D' a deterministic distribution, and 'W' a Weibull distribution with appropriate parameters) and labelling L. The total holding time distributions can be computed from the matrices. For instance

$$\mathbf{H}(1,t) = \begin{cases} 0 & \text{if } t \leq 2, \\ 0.23\,t - 0.46 & \text{if } 2 < t \leq 3, \\ 0.31\,t - 0.7 & \text{if } 3 < t \leq 5, \\ 0.08\,t - 0.24 & \text{if } 5 < t \leq 7, \\ 1 & \text{otherwise.} \end{cases}$$

To describe how the system evolves from state to state, suppose that the boilers starts in state 0. Matrix $\mathbf{P}$ immediately determines the probability to move to a next state. State 2 is chosen, for instance with probability $\mathbf{P}(0,2) = 0.1$. In this case a sample is immediately drawn from distribution $\mathbf{Q}(0,2,t) = 1 - e^{-0.7\,t}$, say 5.3. The system thus holds state 0 for 5.3 time units before moving to state 2. In state 2, again matrix $\mathbf{P}$ is used to determine the next successor, say state 0, whence a random sample is drawn from the distribution $\mathbf{Q}(2,0)$ to determine the holding time in state 2 before moving back to state 0.   □

Paths. Let $\mathcal{M} = (S, \mathbf{P}, \mathbf{Q}, L)$ be a SMC. A sequence $s_0 \xrightarrow{t_0} s_1 \xrightarrow{t_1} s_2 \xrightarrow{t_2} \ldots$, with $s_i \in S$ and $t_i \in \mathbb{R}_{\geq 0}$ such that $\mathbf{P}(s_i, s_{i+1}) > 0$ for all i, is called a *path* through $\mathcal{M}$. For path σ and $i \in \mathbb{N}$, let $\sigma[i] = s_i$, the $(i+1)$-st state of σ, and $\delta(\sigma, i) = t_i$, the time spent in s_i. For $t \in \mathbb{R}_{\geq 0}$ and i the smallest index with $t \leq \sum_{j=0}^{i} t_j$ let $\sigma@t = \sigma[i]$, the state in σ occupied at time t.

Let $Path^{\mathcal{M}}$ denote the set of paths in the SMC $\mathcal{M}$, and $Path^{\mathcal{M}}(s)$ the set of paths in $\mathcal{M}$ that start in s. The superscript $\mathcal{M}$ is omitted unless needed for distinction purposes.

Borel space. A probability measure Pr on sets of paths through a SMC is defined using the standard cylinder construction as follows. Let $s_0, \ldots, s_k \in S$ with $\mathbf{P}(s_i, s_{i+1}) > 0$, $(0 \leq i < k)$, and $I_0, \ldots, I_{k-1}$ non-empty intervals in $\mathbb{R}_{\geq 0}$. Then, $C(s_0, I_0, \ldots, I_{k-1}, s_k)$ denotes the *cylinder set* consisting of all paths $\sigma \in Path(s_0)$ such that $\sigma[i] = s_i$ $(i \leq k)$, and $\delta(\sigma, i) \in I_i$ $(i < k)$. Let $\mathcal{F}(Path)$ be the smallest σ-algebra on $Path$ which contains all sets $C(s, I_0, \ldots, I_{k-1}, s_k)$ where $s_0, \ldots, s_k$ ranges over all state-sequences with $s = s_0$, $\mathbf{P}(s_i, s_{i+1}) > 0$ $(0 \leq i < k)$, and $I_0, \ldots, I_{k-1}$ ranges over all sequences of non-empty intervals in $\mathbb{R}_{\geq 0}$. The probability measure Pr on $\mathcal{F}(Path(s))$ is the unique measure defined by induction on k by $\Pr(C(s_0)) = 1$ and for $k \geq 0$:

$$\Pr(C(s_0, I_0, \ldots, s_k, I', s')) = \Pr(C(s_0, I_0, \ldots, s_k)) \cdot$$
$$\mathbf{P}(s_k, s') \cdot (\mathbf{Q}(s_k, s', b) - \mathbf{Q}(s_k, s', a))$$

where $a = \inf I'$ and $b = \sup I'$. With this definition, a path $s_0 \xrightarrow{t_0} s_1 \xrightarrow{t_1} s_2 \ldots$ corresponds to a sequence $(s_0, 0), (s_1, t_0), (s_2, t_0 + t_1), \ldots$ of bivariate random variables satisfying the properties of Markov renewal sequences [18]. This observation links our definition of SMCs to the standard definition found in the literature.

On the basis of the probability measure Pr, we can define various measures determining the behaviour of a SMC as time passes. For instance,

$$\pi(s, s', t) = \Pr\{\sigma \in Path(s) \mid \sigma@t = s'\}$$

defines the probability distribution on S (ranged over by s') at time t if starting in state s at time 0. We are particularly interested in two specific measures discussed below.

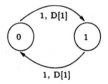

Fig. 2. A SMC without steady-state.

First passage time analysis. We are interested in a measure that describes the probability

$$F(s, s', t) = \Pr\{\sigma \in Path(s) \mid \exists t' \in [\delta(\sigma, 0), t] . \sigma @ t' = s'\}$$

of reaching state s' for the first time within t time units when starting in state s. Note that even if $s = s'$ only paths are considered that leave the state s, since t' has to be at least $\delta(\sigma, 0)$ which is the time needed to leave s. From [18] we have that $F(s, s', t)$ (with $s, s' \in S$) satisfies the following system of equations:

$$F(s, s', t) = \mathbf{P}(s, s') \, \mathbf{Q}(s, s', t) + \sum_{s'' \neq s'} \int_0^t \mathbf{P}(s, s') \, \frac{d\mathbf{Q}(s, s'', x)}{dx} \, F(s'', s', t-x) \, dx$$

Intuitively, the probability to reach state s' from state s for the first time within t time units equals the sum of the probability of taking a direct transition from s to s' (within t time units) and the probability of moving via some intermediate state s'' at time x, yet reaching state s' in the remaining time interval $t - x$. It can be proven that this equation system has a unique solution if the state holding time for any state in the SMC is positive with nonzero probability (as we have assumed) [18].

Long-run average analysis. The long-run average behaviour of a SMC is not as homogeneous as it is for CTMCs. In particular the steady-state behaviour (usually defined as the limit of $\pi(s, s', t)$ for $t \to \infty$) may not exist.

Example 2. Consider for instance, the SMC depicted in Figure 2. For any $t \geq 0$ the probability $\pi(s, s', t)$ does not equal $\pi(s, s', t + 1)$, because the probability mass alternates between the two states. Thus, a limit for $t \to \infty$ of $\pi(s, s', t)$ does not exist. □

However, we can define a related measure based on the average amount of time spent in some state, similar to [12]. For this purpose, we fix a state s, and let σ_s be a path taken randomly from the set $Path(s)$. Then, the quantity $\mathbf{1}_{s'}(\sigma_s @ t)$ is a random variable, indicating whether the state s' is occupied at time t when starting in s. Here we use the characteristic function $\mathbf{1}_{s'}(s'') = 1$ if $s' = s''$ and 0 otherwise.

On the basis of this, we can define a random variable that cumulates the time spent in some state s' up to time t (starting in s) by $\int_0^t \mathbf{1}_{s'}(\sigma_s @ x) \, dx$, and normalise it by the time t in order to obtain a measure of the fraction of time spent in state s' up to time t. Since this is still a random variable, we can derive its expected value. This value corresponds to the average fraction of time spent in state s' in the time frame up to t. For the long-run average fraction of time, we consider the limit $t \to \infty$, as in [23].

Definition 1. *The average fraction of time $T(s, s')$ spent in state s' on the long run when starting in state s is given by:*

$$T(s, s') = \lim_{t \to \infty} E \left[\frac{1}{t} \int_0^t 1_{s'}(\sigma_s @x) \, dx \right]$$

where σ_s ranges randomly over Path(s).

This measure exists for SMCs whenever the expected values of all the distributions $Q(s, s')$ are finite (as we have assumed). Note that for finite CTMCs the measure $T(s, s')$ agrees with the usual steady-state limit $\lim_{t \to \infty} \pi(s, s', t)$. In this sense, T conservatively extends the steady-state measure of CTMCs.

3   CSL on Semi-Markov Chains

This section recalls the syntax of the continuous stochastic logic CSL, and defines its semantics in terms of semi-Markov chains.

Syntax. CSL is a branching-time temporal logic à la CTL [9] with state- and path-formulas based on [5,4].

Definition 2. *Let $p \in [0,1]$, $\trianglelefteq \in \{\leq, \geq\}$, $t \in \mathbb{R}_{\geq 0}$, and $a \in AP$. The syntax of CSL state-formulas is defined by the following grammar:*

$$\Phi ::= \textbf{true} \quad \Big| \quad a \quad \Big| \quad \Phi \wedge \Phi \quad \Big| \quad \neg \Phi \quad \Big| \quad S_{\trianglelefteq p}(\Phi) \quad \Big| \quad \mathcal{P}_{\trianglelefteq p}(\varphi)$$

where for $t \in \mathbb{R}_{\geq 0}$ path-formulas are defined by

$$\varphi ::= X\Phi \quad \Big| \quad \Phi \mathcal{U} \Phi \quad \Big| \quad \Phi \mathcal{U}^{\leq t} \Phi.$$

Other boolean connectives are derived in the usual way, i.e. **false** $= \neg$**true**, $\Phi_1 \vee \Phi_2 = \neg(\neg \Phi_1 \wedge \neg \Phi_2)$, and $\Phi_1 \to \Phi_2 = \neg \Phi_1 \vee \Phi_2$.

The intended meaning of the temporal operators $\mathcal{U}$ ("until") and X ("next step") is standard. We recall from [5] the intuitive meaning of $\mathcal{U}^{\leq t}$, $\mathcal{P}$ and S: The path-formula $\Phi_1 \mathcal{U}^{\leq t} \Phi_2$ is satisfied iff there is some $x \in [0, t]$ such that Φ_1 continuously holds during the interval $[0, x[$ and Φ_2 becomes true at time instant x. $\mathcal{P}_{\trianglelefteq p}(\varphi)$ asserts that the probability measure of the paths satisfying φ falls in the interval $I_{\trianglelefteq p}$. The state formula $S_{\trianglelefteq p}(\Phi)$ asserts that the long-run average fraction of time for a Φ-state falls in the interval $I_{\trianglelefteq p} = \{q \in [0,1] \mid q \trianglelefteq p\}$. Temporal operators like $\Diamond$, $\Box$ and their real-time variants $\Diamond^{\leq t}$ or $\Box^{\leq t}$ can be derived, e.g. $\mathcal{P}_{\trianglelefteq p}(\Diamond^{\leq t} \Phi) = \mathcal{P}_{\trianglelefteq p}(\textbf{true} \, \mathcal{U}^{\leq t} \Phi)$ and $\mathcal{P}_{\trianglerighteq p}(\Box \Phi) = \mathcal{P}_{\trianglelefteq 1-p}(\Diamond \neg \Phi)$.

Semantics. The state-formulas are interpreted over the states of a SMC. Let $\mathcal{M} = (S, \mathbf{P}, \mathbf{Q}, L)$ with proposition labels in AP. The definition of the satisfaction relation $\models \subseteq S \times$ CSL is as follows. Let $Sat(\Phi) = \{\, s \in S \mid s \models \Phi \,\}$.

$$s \models \textbf{true} \text{ for all } s \in S, \qquad s \models \Phi_1 \wedge \Phi_2 \text{ iff } s \models \Phi_i, i \in \{1, 2\},$$
$$s \models a \qquad \text{iff } a \in L(s), \qquad s \models \mathcal{S}_{\unlhd p}(\Phi) \text{ iff } T_{Sat(\Phi)}(s) \in I_{\unlhd p},$$
$$s \models \neg\Phi \qquad \text{iff } s \not\models \Phi, \qquad s \models \mathcal{P}_{\unlhd p}(\varphi) \text{ iff } Prob(s, \varphi) \in I_{\unlhd p}.$$

Here, $T_{S'}(s)$ denotes the average fraction of time spent in $S' \subseteq S$ with respect to state s, i.e.

$$T_{S'}(s) = \sum_{s' \in S'} T(s, s').$$

Recall that $T(s, s')$ conservatively extends the definition of a steady-state distribution for CTMCs. $Prob(s, \varphi)$ denotes the probability measure of all paths $\sigma \in Path(s)$ satisfying φ, i.e.

$$Prob(s, \varphi) = \Pr\{\, \sigma \in Path(s) \mid \sigma \models \varphi \,\}.$$

The fact that, for each state s, the set $\{\, \sigma \in Path(s) \mid \sigma \models \varphi \,\}$ is measurable, follows by easy verification. The satisfaction relation (also denoted $\models$) for the path-formulas is defined as usual:

$$\sigma \models X\Phi \qquad \text{iff } \sigma[1] \text{ is defined and } \sigma[1] \models \Phi,$$
$$\sigma \models \Phi_1 \mathcal{U} \Phi_2 \quad \text{iff } \exists k \geq 0. \, (\sigma[k] \models \Phi_2 \wedge \forall 0 \leq i < k. \, \sigma[i] \models \Phi_1),$$
$$\sigma \models \Phi_1 \mathcal{U}^{\leq t} \Phi_2 \text{ iff } \exists x \in [0, t]. \, (\sigma@x \models \Phi_2 \wedge \forall y \in [0, x[. \, \sigma@y \models \Phi_1).$$

4   Model Checking SMCs against CSL

Model checking SMCs against CSL follows the usual strategy: Given a model $\mathcal{M} = (S, \mathbf{P}, \mathbf{Q}, L)$ and a state-formula Φ, the set $Sat(\Psi)$ is recursively computed for the sub-formulas of Φ. This can proceed via well studied means [16,5] (on the embedded DTMC $(S, \mathbf{P}, L)$) except for the time-bounded until operator $\mathcal{U}^{\leq t}$, and for the long-run operator $\mathcal{S}$. These two operators require specific care.

Time-bounded until. For computing the probability of satisfying a time-bounded until formula, we closely follow the strategy of [6], and reduce the problem to a well studied transient measure. More precisely, it will turn out that we can compute the time-bounded until probabilities via a first passage time analysis in a derived SMC, where certain subsets of states are made absorbing. To this end, we let $\mathcal{M}[\Phi]$ (for SMC $\mathcal{M}$ and state formula Φ) denote the SMC obtained from $\mathcal{M}$ by making all Φ-states absorbing. We have:

Theorem 1. *Let* $\mathcal{M} = (S, \mathbf{P}, \mathbf{Q}, L)$ *be a SMC, and* Φ_1 *and* Φ_2 *be CSL state-formulas. Then*

$$Prob^{\mathcal{M}}(s, \Phi_1 \mathcal{U}^{\leq t} \Phi_2) = Prob^{\mathcal{M}[\neg\Phi_1 \vee \Phi_2]}(s, \Diamond^{\leq t} \Phi_2)$$
$$= \begin{cases} 1 & \text{if } s \models \Phi_2, \\ \sum_{s' \models \Phi_2} F^{\mathcal{M}[\neg\Phi_1 \vee \Phi_2]}(s, s', t) & \text{otherwise.} \end{cases}$$

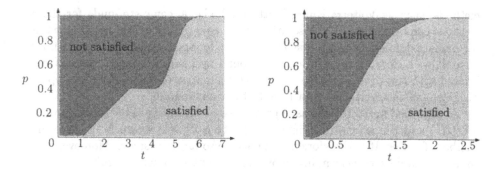

Fig. 3. Satisfaction of $\mathcal{P}_{\geq p}(\text{leaking}\,\mathcal{U}^{\leq t}\text{working})$ for state 2 and 3 of the boiler example.

Proof: The proof of the first equality is based on a bijection between the paths in $\mathcal{M}$ satisfying $\Phi_1\mathcal{U}^{\leq t}\Phi_2$ and the paths in $\mathcal{M}[\neg\Phi_1 \vee \Phi_2]$ satisfying $\Diamond^{\leq t}\Phi_2$, up to the state where Φ_2 becomes satisfied, and hence over the whole path-prefixes contributing to the two probability measures $\Pr\{\sigma \in Path^{\mathcal{M}} \mid \sigma \models \Phi_1\mathcal{U}^{\leq t}\Phi_2\}$ and $\Pr\{\sigma \in Path^{\mathcal{M}[\neg\Phi_1\vee\Phi_2]} \mid \sigma \models \Diamond^{\leq t}\Phi_2\}$. With respect to the second equality we only consider the case $s \not\models \Phi_2$. In this case $\sigma \models \Diamond^{\leq t}\Phi_2$ can be shown to hold if and only if $\exists t' \in [\delta(\sigma,0), t]\,.\,\sigma@t' \models \Phi_2$, since $\sigma[0] \not\models \Phi_2$. The proof follows from the definition of F and the fact that Φ_2-states are absorbing, justifying the summation over all Φ_2-states.   □

Example 3. Returning to the boiler example of Fig, 1, let us check the time-bounded until formula $\mathcal{P}_{\geq p}(\text{leaking}\,\mathcal{U}^{\leq t}\text{working})$. First, we observe that state 1 does neither satisfy leaking nor working, and hence state 1 does not satisfy the path-formula leaking $\mathcal{U}^{\leq t}$working with positive probability. In contrast, according to Theorem 1, state 0 satisfies the path formula with probability 1, because $0 \models$ working.

The remaining states 2 and 3 are more interesting. Following Theorem 1 we need to investigate a SMC where state 0 and 1 are made absorbing, and compute the probability of satisfying leaking $\mathcal{U}^{\leq t}$working via the values of $F(2,0,t)$, respectively $F(3,0,t)$ in this SMC. The values of these functions are plotted in Fig. 3. One can see that for pairs (p,t) above the plot the formula is invalid, while it is valid for pairs below the plot (and for the plot itself).   □

While in the above example the values of F can be calculated directly, the situation is more involved in general. Recall that $F(s, s', t)$ is the unique solution of the system of equations

$$F(s, s', t) = \mathbf{P}(s, s')\,\mathbf{Q}(s, s', t) + \sum_{s'' \neq s'} \int_0^t \mathbf{P}(s, s')\frac{d\mathbf{Q}(s, s'', x)}{dx}\,F(s'', s', t-x)\,dx.$$

This system of equations can be classified as a system of Volterra equations of the second type. In principle it is possible to solve them by appropriate numerical methods, such as Volterra-Runge-Kutta methods. A complete guide for these methods can be found in [7]. A solution of the equations can also be obtained in the Laplace domain. This approach works good for small systems and sometimes even allows a closed-form solution to be found by hand. For larger systems one is faced with two problems. One has to invert a matrix of functions in a complex variable, and to reverse the transform to the time domain.

As described in [14,15], the asymptotic space complexity of the latter method is $\mathcal{O}(N^2)$ and the asymptotic time complexity is $\mathcal{O}(N^4)$ where $N = |S|$ is the number of states. It is therefore not applicable to larger systems. Moreover, the numerical Laplace transform inversion can encounter numerical problems under some conditions. It is also possible to solve this Volterra system by transforming it to a system of partial differential equation, a system of ordinary differential equation, initial and boundary conditions, and a system of integral equations. [14] contains a comparison of these two approaches together with numerical considerations.

Long-run average. For model checking the operator $S_{\unlhd p}(\Phi)$ one needs to accumulate the average fraction of time quantities $T(s, s')$ for each state s' satisfying Φ. If $\mathcal{M}$ is a strongly connected[1] SMC, $T(s, s')$ can be obtained via the equilibrium probability vector π of the embedded DTMC $(S, \mathbf{P}, L)$, which in turn is given as the unique solution of the linear equation system

$$\pi(s) = \sum_{s' \in S} \mathbf{P}(s', s) \cdot \pi(s') \text{ such that } \sum_{s \in S} \pi(s) = 1.$$

Theorem 2. *[8] Let $\mathcal{M} = (S, \mathbf{P}, \mathbf{Q}, L)$ be a strongly connected SMC, and π be as above. Then*

$$T(s, s') = \frac{\pi(s')\mu(s')}{\sum_{s'' \in S} \mu(s'')\pi(s'')}$$

where $\mu(s'')$ is the expected holding time in state, i.e., $\mu(s'') = E[\mathbf{H}(s'')]$.

Notice that $T(s, s')$ is independent of the starting state s in this case. If otherwise $\mathcal{M}$ is not strongly connected, we proceed as in [5], and isolate the bottom strongly connected subsets of S via a graph algorithm [22]. Whenever state s' is not a member of any bottom strongly connected subset of S, we have $T(s, s') = 0$. The following result allows model checking the S operator in the other cases. We let $Prob(s, \Diamond B)$ denote the probability of eventually reaching the set $B \subseteq S$ from state s. This quantity can be computed via the embedded DTMC $(S, \mathbf{P}, L)$ [16].

Theorem 3. *Let $\mathcal{M} = (S, \mathbf{P}, \mathbf{Q}, L)$ be a SMC, B a bottom strongly connected subset of S, and $s' \in B$. Then:*

$$T(s, s') = Prob(s, \Diamond\, B) \cdot T^B(s', s')$$

where the superscript B refers to the strongly connected SMC $\mathcal{M}^B$ spanned by B.

---

[1] A SMC is strongly connected if there is some k such that $\mathbf{P}^k(s, s') > 0$ for each s, s'.

Proof: We only consider the case where $s \notin B$ can reach B with positive probability. The idea of the proof is to count the average time that the SMC $\mathcal{M}$ spends in class B. Once we have isolated this quantity we are able to compute the fraction of time $\mathcal{M}$ spends in a particular state of this class. Let $1_B(s') = 1$ if $s' \in B$ and 0 otherwise. We shall calculate the exact value for

$$E\left[\frac{1}{t}\int_0^t 1_B(\sigma_s @x)\,dx\right]$$

where σ_s ranges over $Path(s)$. Let $\tilde{t}$ be the time of absorption in B (if σ_s touches B otherwise $\tilde{t} = \infty$), $\tilde{t}$ is be a random variable and depends on the path σ_s drawn from $Path(s)$. The distribution of $\tilde{t}$ is given by $\Pr\{\tilde{t} \le t'\} = F(s, B, t') = \sum_{s' \in B} F(s, s', t')$ where the latter is the first passage time distribution mentioned earlier.

Since B is bottom strongly connected, the function $1_B(\sigma_s @x)$ will be constant 1 from $\tilde{t}$ on. So, for $t \ge \tilde{t}$ we have that

$$\int_0^t 1_B(\sigma_s @x)\,dx = \frac{t - \tilde{t}}{t}$$

and otherwise (i.e., $t < \tilde{t}$) the integral equals 0. So, for fixed t the above integral describes a random variable R_t as follows:

$$R_t(\tilde{t}) = \begin{cases} \frac{t - \tilde{t}}{t} & \text{if } t \ge \tilde{t}, \\ 0 & \text{otherwise.} \end{cases}$$

The distribution of R_t is

$$\Pr\{R_t \le x\} = \Pr\{(t - \tilde{t})/t \le x\} + \Pr\{R_t = 0\}$$

which can be rewritten, using that $F(s, B, x)$ is the distribution of $\tilde{t}$, to

$$\Pr\{R_t \le x\} = 1 - F(s, B, t - xt) + \Pr\{R_t = 0\}.$$

Now, the expected value $E[R_t]$ is obviously

$$\int_0^1 u\,\frac{d(1 - F(s, B, t - u\,t))}{du}\,du + 0\,\frac{d(\Pr\{R_t = 0\})}{du} = \int_0^1 u\,t\,\frac{dF(s, B, t - u\,t)}{du}\,du.$$

Substituting $u = \frac{t - y}{t}$ we get

$$E[R_t] = \frac{1}{t}\int_0^t (t - y)\,\frac{dF(s, B, y)}{dy}\,dy.$$

What we are looking for is the limit of this quantity as $t \to \infty$ given by

$$\lim_{t\to\infty} F(s, B, t) - \lim_{t\to\infty} \frac{1}{t}\int_0^t y\,\frac{dF(s, B, y)}{dy}\,dy = Prob(s, \Diamond B) - \lim_{t\to\infty} \frac{E[\tilde{t}]}{t}$$

where $E[\tilde{t}]$ is the expected value of $\tilde{t}$. Recall that $\tilde{t}$ is distributed according to $F(s, B, t)$. Also note that $\lim_{t\to\infty} F(s, B, t) = Prob(s, \Diamond B)$. Since we have assumed that state s can reach B with positive probability (and all distributions have finite means) $E[\tilde{t}]$ needs to be finite and hence

$$\lim_{t\to\infty} E\left[\frac{1}{t}\int_0^t \mathbf{1}_B(\sigma_s @ x)\,dx\right] = Prob(s, \Diamond B).$$

The proof of the theorem follows from this result by two observations. First, the time of entering B is a renewal point, i.e., a time instant where the future behaviour of the stochastic process does only depend on the currently occupied state. Second, the fraction of time spent in a particular state inside B is independent of the starting state – due to strong connectedness – if assuming to start inside B.   □

Example 4. Let us check a long-run average property for the example boiler system, such as $S_{\leq p}(\text{working})$. We first observe that the SMC in Fig. 1 is strongly connected. Theorem 2 requires the computation of the expected holding times for each state of the SMC, resulting from weighted sums of the involved distributions. We get $\mu(0) = 70.319$, $\mu(1) = 3.95$, $\mu(2) = 3.2$, and $\mu(3) = 0.887$. Next, we solve the embedded DTMC, and obtain a vector $\pi = [0.686, 0.1373, 0.109, 0.065]$. Finally we compute $T_{\text{Sat(working)}}(s) = 0.981$. Since the SMC is strongly connected, this value is independent of the state s chosen, and hence $S_{\leq p}(\text{working})$ is satisfied (for all states) whenever $0.981 \leq p$.   □

Apart from the need to derive expected values of general distributed random variables, the numerical algorithms needed for model checking the long-run average operator are the same as the ones needed for checking CTMCs [5].

5   Concluding Remarks

In this paper, we investigated adapting CSL model checking to semi-Markov chains, an extension of CTMCs in which state holding times are governed by general distributions. To achieve a smooth extension of the theory we developed an enhanced definition of long-run properties and proved novel results required for model checking not strongly connected SMCs. On the practical side, the conclusion we draw from our investigation is partially negative: verifying a CSL-formula can become numerically very complex when dropping the memoryless property. This is caused by the involved procedure needed for checking time-bounded formulas such as timed probabilistic reachability properties. We proved that long-run properties and (untimed) eventualities can be checked without an increase in complexity compared to the CTMC case, though.

The SMC model considered in this paper incorporates general distributions, but is known to be of limited use to model concurrent delays. Compositional extensions of SMCs – such as generalised semi-Markov chains or stochastic automata [11]– are more elegant to apply in this context. It is worth to highlight that our practically negative result concerning the model checking of time-bounded formulas carries over to these models. Further research is needed to investigate whether abstraction techniques or weaker temporal properties – like expected time properties – yield a practical solution for such models.

References

1. R. Alur, C. Courcoubetis and D. Dill. Model-checking in dense real-time. *Inf. and Comp.*, **104**: 2–34, 1993.
2. S. Asmussen, O. Nermand and M. Olsson. Fitting phase-type distributions via the EM algorithm. *Scand. J. Statist.*, **23**: 420–441, 1996.
3. A. Aziz, K. Sanwal, V. Singhal and R. Brayton. Verifying continuous time Markov chains. In R. Alur and T.A. Henzinger (eds), *Computer-Aided Verification*, LNCS 1102, pp. 269–276, Springer, 1996.
4. A. Aziz, K. Sanwal, V. Singhal and R. Brayton. Model checking continuous time Markov chains. *ACM Trans. on Computational Logic*, **1**(1): 162–170, 2000.
5. C. Baier, J.-P. Katoen and H. Hermanns. Approximate symbolic model checking of continuous-time Markov chains. In J.C.M. Baeten and S. Mauw (eds), *Concurrency Theory*, LNCS 1664, pp. 146–162, Springer, 1999.
6. C. Baier, B. R. Haverkort, H. Hermanns and J.-P. Katoen. Model checking continuous-time Markov chains by transient analysis. In E.A Emerson and A.P. Sistla (eds), *Computer Aided Verification*, LNCS 1855, pp. 358–372, Springer, 2000.
7. H. Brunner and P. van der Houwen, *The Numerical Solution of Volterra Equations*. North Holland, 1986.
8. E. Cinlar. *Introduction to Stochastic Processes*. Prentice-Hall Inc., 1975.
9. E. Clarke, E. Emerson and A. Sistla. Automatic verification of finite-state concurrent systems using temporal logic specifications. *ACM Transactions on Programming Languages and Systems*, **8**: 244–263, 1986.
10. M.E. Crovella. Performance evaluation with heavy tailed distributions (extended abstract). In B. Haverkort, H. Bohnenkamp and C. Smith (eds), *Computer Performance Evaluation*, LNCS 1786, pp. 1–9, Springer, 2000.
11. P.R. D'Argenio, J.-P. Katoen and E. Brinksma. An algebraic approach to the specification of stochastic systems (extended abstract). In D. Gries and W.-P. de Roever (eds), *Programming Concepts and Methods*. Chapman & Hall, pp. 126–147, 1998.
12. L. de Alfaro. *Formal Verification of Probabilistic Systems*. PhD thesis, Stanford University, 1997.
13. A. Feyzi Ates, M. Bilgic, S. Saito, and B. Sarikaya. Using timed CSP for specification, verification and simulation of multimedia synchronization. *IEEE J. on Sel. Areas in Comm.*, **14**:126–137, 1996.
14. R. German *Performance Analysis of Communication Systems: Modeling with Non-Markovian Stochastic Petri Nets*. John Wiley & Sons, 2000.
15. R. German, D. Logothe, and K.S. Trivedi. Transient analysis of Markov regenerative stochastic Petri nets: A comparison of approaches. *Proc. 6th Int. Workshop on Petri Nets and Performance Models*, pages 103–112, IEEE CS Press, 1995.
16. H. Hansson and B. Jonsson. A logic for reasoning about time and reliability. *Formal Aspects of Computing* **6**: 512–535, 1994.
17. D. Van Hung and Z. Chaochen. Probabilistic duration calculus for continuous time. *Formal Aspects of Computing*, **11**: 21–44, 1999.
18. V. Kulkarni. *Modeling and Analysis of Stochastic Systems*. Chapman & Hall, 1995.

19. M.Z. Kwiatkowska, G. Norman, R. Segala, and J. Sproston. Verifying quantitative properties of continuous probabilistic timed automata. In C. Palamadessi (ed), *Concurrency Theory*, LNCS 1877, pp. 123-137, Springer, 2000.

20. W. Nelson. Weibull analysis of reliability data with few or no failures. *Journal of Quality Technology* **17** (3), 140-146, 1985.

21. A.D. Swain and H.E. Guttmann. Handbook of human reliability analysis with emphasis on nuclear power plant applications - final report. Technical Report NRC FIN A 1188 NUREG/CR-1278 SAND80-0200, US Nuclear Regulatory Commission, 1983.

22. R.E. Tarjan. Depth-first search and linear graph algorithms. *SIAM Journal of Computing*, 1: 146-160, 1972.

23. H. Taylor and S. Karlin. *An Introduction To Stochastic Modeling*. Academic Press, 1998.

Coin Lemmas with Random Variables

Katia Folegati and Roberto Segala

Department of Computer Science
University of Bologna - Italy

Abstract. Coin lemmas are a tool for decoupling probabilistic and non-deterministic arguments in the analysis of concurrent probabilistic systems. They have revealed to be fundamental in the analysis of randomized distributed algorithms, where the interplay between probability and nondeterminism has proved to be subtle and difficult to handle.
We reformulate coin lemmas in terms of random variables obtaining a new collection of coin lemmas that is independent of the underlying computational model and of more general applicability to the study of concurrent nondeterministic probabilistic systems.

1 Introduction

Coin lemmas are a tool for decoupling probabilistic and nondeterministic arguments in the analysis of concurrent probabilistic systems [12]. They have revealed to be fundamental in the analysis of randomized distributed algorithms [7, 8], where the interplay between probability and nondeterminism has proved to be subtle and difficult to handle [10].

Coin lemmas are formulated in the framework of probabilistic automata [11], a probabilistic extension of labeled transition systems (automata) where the notion of transition is extended so that a transition from some state s leads to a discrete probability distribution over states rather than to a single state. Each state enables several transitions, and the choice of which transitions to schedule is left unspecified. Nondeterminism is resolved by an entity called a *scheduler*, and the result of resolving nondeterminism can be described as an acyclic Markov process, called a *probabilistic execution*. Properties that involve probability can be studied on probabilistic executions, while typical arguments about probabilistic automata involve proving upper and lower bounds on the probabilities of events under the action of any scheduler, the events being sets of paths in the Markov process that represents a probabilistic execution.

A coin lemma is a tool to prove upper and lower bounds on probabilities of events within any probabilistic execution. Specifically, a coin lemma provides us with a rule to associate an event with each probabilistic execution together with a lower bound on the probability of the associated events. Typically the events associated with a probabilistic execution are derived from events in a known stochastic process, and similarly, the lower bounds on probabilities are derived from the known stochastic process. The analysis of a concurrent probabilistic system is then reduced to the analysis of the events returned by the rules, which, being just ordinary sets of paths, do not include any probability.

L. de Alfaro and S. Gilmore (Eds.): PAPM-PROBMIV 2001, LNCS 2165, pp. 71–86, 2001.

1.1   Coin Lemmas in the Setting of Algorithms

To understand better the role of coin lemmas we illustrate what happens in the area of distributed algorithms, which is the area where coin lemmas were proposed first. When studying a distributed algorithm the objective is to state that for every scheduler (the entity that decides the temporal ordering of the events at different processors) the probability of successful termination is at least some number p. That is, translated in terms of probabilistic automata, for each probabilistic execution the probability of the event that expresses successful termination is at least p.

In practice it is more convenient to show that a sub-event of successful termination has probability at least p, the sub-event expressing the fact that some random draws that occur in the algorithm give some specific results.

Example 1. The randomized algorithm of Lehmann and Rabin for the dining philosophers problem [6] works as follows: whenever a philosopher wants to eat he/she flips a coin to decide which fork to pick up first, waits for the fork to be free and picks it up, and finally checks whether the other fork is free. If the other fork is free, then the philosopher picks it up and eats, otherwise the philosopher puts down the first fork and starts again. It turns out that if two neighbor philosophers draw opposite coins at a certain stage, then the algorithm terminates successfully and one philosopher eats. However, the algorithm may terminate successfully even if the coins are not opposite. Studying the sub-event of successful termination that states that two coins are opposite is enough anyway to state that there is a high chance that some philosophers eat eventually.

In summary, the designer of an algorithm thinks of a stochastic process that supposedly occurs in any probabilistic execution, thinks of the fact that whenever the stochastic process gives some specific results R the algorithm terminates successfully, and concludes that the algorithm terminates successfully with a probability that is as high as the probability of R. Unfortunately, in most of the cases the stochastic process that the designer has in mind does not take place since, for example, the algorithm may terminate before the stochastic process completes.

Since designers rarely consider explicitly the fact that the stochastic process they have in mind may not complete, they often come to the wrong conclusion that an algorithm is correct whenever it is possible that the stochastic process does not complete and at the same time the algorithm does not terminate successfully.

Example 2. Several randomized algorithms in the literature are based on the following argument. The processors communicate with each other and at the end of the communication it turns out that some of them have participated in a game. If there is a unique winner in the game, then the algorithm terminates. The game consists of flipping fair coins until a head comes out and counting how many coins are flipped. Thus, the probability of drawing number i is $1/2^i$. The winner is the player who draws the highest number. We know that, given that there are k players, the probability of a unique winner is at least some constant c independent of k. Should we conclude that the algorithm terminates successfully with probability at least c?

Although in the literature we can find statements like the one above, the correctness of the statement depends on how the number of players is chosen. If the players are chosen before starting the game, then the lower bound c does hold. Otherwise, we may add players to the game until we obtain two winners, which would violate considerably the lower bound c.

Coin lemmas have the scope of formalizing the arguments that designers usually put forward about their algorithms and of avoiding the pitfalls given by incomplete stochastic processes. Thus, given a stochastic process, a coin lemma provides a mechanical rule to pick up an event in each probabilistic execution, the event representing the successful part of the stochastic process, and provides a lower bound on the probabilities of the chosen events that is the same as the probability of the successful part of the stochastic process. The main problem is to decide what event to return whenever the stochastic process does not complete in a probabilistic execution.

Viewing a stochastic process as a sequence of elementary experiments, the rule of a coin lemma identifies all the experiments that do take place in an execution (a path of the probabilistic execution) and their results. Each path in a probabilistic execution appears in the event returned by the rule if it is possible to fix the outcomes of the experiments that do not take place so that the underlying stochastic process is successful. This idea is called "the essence of a coin lemma" in [12], since it is the common denominator of all the coin lemmas that have been proposed so far.

The advantage of this formulation of coin lemmas is that the designer of an algorithm is given a set of paths (ordinary executions) that must be shown to lead to successful termination if he/she wants to conclude that the algorithm is successful with some minimum probability. The executions can be analyzed without resorting to probabilities. Furthermore, the rule would say something like "if a certain coin is not flipped then the resulting execution must be shown to lead to successful termination". Thus, the designer of the algorithm will be forced to prove either that the coin is always flipped or that the algorithm is indeed successful whenever the coin is not flipped.

Example 3. The coin lemma that we should use for the argument of Example 2 would be associated with the stochastic process where k numbers are drawn. Assuming that we have some way to decide which k to use, the scheduler that adds players until there are two winners would be discovered immediately since in the event returned by the rule we would have several executions where less than k numbers are drawn.

1.2 Our Contribution

A weak point in the current formulations of coin lemmas [12] is that the rules are formulated in a very restrictive language. As a result, many times simple and obvious variations to the rules require the proof of a new coin lemma. In this paper we propose a new formulation of coin lemmas based on random variables that gives us much more flexibility in the formulation of a rule without forcing us to prove new coin lemmas each time. Another advantage of the new formulation is that the concept of a random variable is independent of the structure of the probability space that represents a probabilistic execution. Thus, the new coin lemmas can be used easily on several models. On the contrary, the current formulation is heavily based on the structure of the transitions of a probabilistic automaton.

If we express a stochastic process as a sequence of elementary random experiments, the main objective for the formulation of a rule is to identify the places in an execution where each experiment takes place and its corresponding outcome. For this reason, we start with the concept of an *experiment*, which is a pair (D, U) where D is a 0/1-valued random variable that identifies the places where the experiment takes place, and U is a random variable that identifies the successful outcomes. An experiment has taken place in an execution α whenever $D(\alpha) = 1$. We define the notion of a p-successful experiment, which means that the probability of success is at least p whenever the experiment takes place. Then, we prove that in each probabilistic execution the probability that either a p-successful experiment does not take place or that the experiment takes place and gives successful results is at least p.

The result about p-successful experiments allows us to capture immediately the coin lemmas of [12] that deal with single occurrences of actions; however, many more general properties can be captured easily like the intricate coin lemma necessary for the proof of correctness of the consensus algorithm of Ben-Or [1] (see [11] for a proof of correctness based on coin lemmas). The identification of the intermediate concept of p-successful experiment and the use of random variables turned out to be a considerable improvement over the formalization of [12].

We elaborate further on the concept of p-successful experiments by showing how several experiments can be combined into a single experiment. In this way we can capture easily the coin lemmas of [12] that look at the outcome of the first experiment that takes place among many experiments.

Finally, we show how to handle more general stochastic processes composed of finite or countable sequences of experiments. We derive two different coin lemmas. In both coin lemmas we consider a collection of experiments $\{(D_i, X_i)\}_{i \in I}$ and a measurable function $f(X_1, X_2, \ldots)$ with values in $\{0, 1\}$. Then we identify the probability distribution of the X_i's under the assumption that all the experiments take place, and we prove a lower bound on the probability of the event that considers all those cases where it is possible to fix the outcomes of the experiments that do not take place so that f evaluates to 1. Once again, if we assume to be successful whenever f evaluates to 1, the principle is that if some experiment does not take place and one of its possible outcomes is considered as successful, then we should make sure that the system under examination behaves correctly.

The rest of the paper is structured as follows. Section 2 gives the necessary background on probability theory and introduces some notation; Section 3 gives a description of probabilistic automata; Section 4 introduces the coin lemmas for single experiments and compares them with the formulation of [12]; Section 5 proves properties of single experiments that allow us to combine several experiments into a unique new experiment; Section 6 introduces the coin lemmas for general stochastic processes; Section 7 gives some concluding remarks.

2   Preliminaries

A σ-field over a set X is a set $\mathcal{F} \subseteq 2^X$ that includes X and is closed under complement and countable union. Observe that 2^X is a σ-field over X. A *measurable space* is a pair $(X, \mathcal{F})$ where X is a set and $\mathcal{F}$ is a σ-field over X. The set X is also called the *sample space*. A measurable space $(X, \mathcal{F})$ is called *discrete* if $\mathcal{F} = 2^X$. In the paper we work mostly with discrete measurable spaces except when we deal with probabilistic executions.

Given a measurable space $(X, \mathcal{F})$, a finite *measure* over $(X, \mathcal{F})$ is a function $\mu : \mathcal{F} \to \mathbb{R}^{\geq 0}$ such that, for each countable collection $\{X_i\}_{i \in I}$ of pairwise disjoint elements of $\mathcal{F}$, $\mu(\cup_I X_i) = \sum_I \mu(X_i)$. For the purpose of this paper we are interested in measures μ where $\mu(X) \leq 1$. A *probability measure* over a measurable space $(X, \mathcal{F})$ is a measure μ over $(X, \mathcal{F})$ such that $\mu(X) = 1$; a *sub-probability measure* over $(X, \mathcal{F})$ is a measure μ over $\mathcal{F}$ such that $\mu(X) \leq 1$. A measure over a discrete measurable space is called a *discrete measure*. We also say that a discrete measure over $(X, 2^X)$ is a discrete measure over X. Sometimes we refer to probability measures as *distributions*.

Given a set X, denote by $Disc(X)$ the set of discrete probability measures on the measurable space $(X, 2^X)$, and denote by $SubDisc(X)$ the set of discrete sub-probability measures on the measurable space $(X, 2^X)$. We call a discrete probability measure a *Dirac* measure if it is concentrated at a single point, say x_0, by assigning measure 1 to any set containing x_0 and measure 0 to any other set. In the paper we use sub-probability measures to denote the fact that sometimes there is no progress from some state s. We view a discrete sub-probability measure μ over X as a discrete probability measure over $X \cup \{\bot\}$ where $\bot$ denotes the fact that there is no progress and $\mu(\bot) = 1 - \mu(X)$ denotes the probability of not making any progress. When $\mu(X) = 0$, then we can think of μ as concentrated at $\bot$. For this reason we abuse notation and say that a discrete sub-probability measure μ over X is Dirac if either μ is a Dirac probability measure or $\mu(\bot) = 1$. We drop the set notation whenever a set is singleton.

A function $f : X_1 \to X_2$ is said to be a *measurable function* from a measurable space $(X_1, \mathcal{F}_1)$ to a measurable space $(X_2, \mathcal{F}_2)$ if for each element C of $\mathcal{F}_2$, $f^{-1}(C) \in \mathcal{F}_1$. In such case, given a measure μ on $(X_1, \mathcal{F}_1)$, we can define the measure *induced* by f, denoted by $f(\mu)$, as $f(\mu)(C) = \mu(f^{-1}(C))$. A *random variable* defined on a measurable space $(X, \mathcal{F})$ is a measurable function from $(X, \mathcal{F})$ to $(\mathbb{R}, B)$, where B denotes the σ-field generated by the open intervals of the reals, also called the *Borel σ-field* over the reals.

3   The Model

In this section we give an overview of probabilistic automata. We omit the distinction between internal and external actions since it is irrelevant for the purpose of this paper. The reader interested in more details is referred to [11].

3.1   Probabilistic Automata

A *probabilistic automaton* $\mathcal{A}$ is a tuple $(S, \bar{s}, \Sigma, \mathcal{D})$ where S is a set of *states*, $\bar{s} \in S$ is a *start state*, Σ is a set of *actions*, and $\mathcal{D} \subseteq S \times \Sigma \times Disc(S)$ is a *transition relation*.

For convenience, states are ranged over by r, q, s; actions are ranged over by a, b, c; discrete distributions are ranged over by μ. We denote the elements of a probabilistic automaton $\mathcal{A}$ by $S, \bar{s}, \Sigma, \mathcal{D}$, and we propagate primes and indicies. Thus, the elements of a probabilistic automaton $\mathcal{A}'_i$ are denoted by $S'_i, \bar{s}'_i, \Sigma'_i, \mathcal{D}'_i$.

We call an element of $\mathcal{D}$ a *transition*. We say that a transition (s, a, μ) is *enabled* from s and is *labeled* by a. We also say that s *enables* (s, a, μ). We call a transition (s, a, μ) *Dirac* if μ is a Dirac measure. For a state s we denote by $\mathcal{D}(s)$ the set of transitions of $\mathcal{D}$ that are enabled from s.

We call a state s *Dirac* if all the transitions enabled from s are Dirac; we call s *deterministic* if it enables at most one transition for each action; we call s *singleton* if it enables at most one transition. We say that a probabilistic automaton $\mathcal{A}$ is *Dirac, deterministic, singleton*, if each state of $\mathcal{A}$ is Dirac, deterministic, singleton, respectively.

For comparison with other models, probabilistic automata can be seen as an extension of ordinary automata (also called labeled transition systems) since an ordinary automaton is essentially a Dirac probabilistic automaton. The reactive systems of [5] are deterministic probabilistic automata; thus, probabilistic automata could be called alternatively *nondeterministic reactive systems*. The probabilistic automata of [9] are equivalent to probabilistic automata. Markov Decision Processes [2] are equivalent to deterministic probabilistic automata. The alternating model of [3] can be seen as a probabilistic automaton where each state is either Dirac or singleton.

3.2   Executions

A *potential execution* of a probabilistic automaton $\mathcal{A}$ is a finite or infinite sequence of alternate states and actions, $\alpha = s_0 a_1 s_1 a_2 s_2 \cdots$, starting from a state and, if the sequence is finite, ending with a state. Define the length of α, denoted by $|\alpha|$, to be the number of occurrences of actions in α. If α is an infinite sequence, then $|\alpha| = \infty$. For a natural number $i \leq |\alpha|$, denote by $\alpha[i]$ the state s_i. In particular, $\alpha[0]$ is the start state of α. If α is finite, then denote by $\alpha[\perp]$ the last state of α.

Two potential executions $\alpha = s_0 a_1 s_1 \cdots a_n s_n$ and $\alpha' = s'_0 a'_1 s'_1 \cdots$ can be concatenated if $s_n = s'_0$. In such case the concatenation, denoted by $\alpha \frown \alpha'$, is the potential execution $s_0 a_1 s_1 \cdots a_n s_n a'_1 s'_1 \cdots$. If $\alpha = \alpha_1 \frown \alpha_2$, then we say that α_1 is a *prefix* of α, denoted by $\alpha_1 \leq \alpha$.

An *execution* of a probabilistic automaton $\mathcal{A}$ is a potential execution of $\mathcal{A}$, $\alpha = s_0 a_1 s_1 a_2 s_2 \cdots$, such that for each $i < |\alpha|$ there exists a transition (s_{i-1}, a_i, μ_i) of $\mathcal{D}$ where $\mu_i(s_i) > 0$. An execution α is said to be *initial* if $\alpha[0] = \bar{s}$. Denote by $execs(\mathcal{A})$ the set of executions of $\mathcal{A}$ and by $iexecs(\mathcal{A})$ the set of initial executions of $\mathcal{A}$. Similarly, denote by $execs^*(\mathcal{A})$ and $iexecs^*(\mathcal{A})$ the set of finite executions and finite initial executions, respectively, of $\mathcal{A}$.

An execution is the result of resolving both nondeterminism and probability in a probabilistic automaton and records the sequence of states that the system goes through, along with the sequence of actions it engages in. In a probabilistic setting it is useful to know the probability distributions over executions that arise after resolving the nondeterminism. This is the argument of the next section.

3.3 Schedulers

A *scheduler* for a probabilistic automaton $\mathcal{A}$ is a function $\sigma : execs^*(\mathcal{A}) \rightarrow SubDisc(\mathcal{D})$ such that for each finite execution α, $\sigma(\alpha) \in SubDisc(\mathcal{D}(\alpha[\bot]))$. We say that a scheduler is *Dirac* if it assigns a Dirac measure to each execution.

In other words, a scheduler is an entity that given a finite execution α chooses arbitrarily either to stop or to perform one of the transitions enabled from the last state of α, possibly using randomization in its choices. Observe that the choice of a scheduler is based on the full history. A scheduler is the entity that resolves the nondeterminism in a probabilistic automaton. In the context of algorithms a scheduler is usually called an *adversary* since it is seen as an entity that tries to degrade the performance of an algorithm as much as possible (the worst case scenario is what matters); in the context of Markov Decision Processes an adversary is called a *policy* since the objective of the area of Markov Decision Processes is to find the best possible strategy to improve performance. What matters from our point of view is that schedulers, adversaries and policies are the same entity, which is the entity that resolves nondeterminism.

In the literature there is a distinction between probabilistic and deterministic schedulers. In this paper we have chosen to use the word scheduler for a probabilistic scheduler and the word Dirac scheduler for a nondeterministic scheduler. The reason for our choice is to avoid overloading the term "nondeterministic".

Consider a scheduler σ and a finite execution α with last state q. The distribution $\sigma(\alpha)$ describes how to move from q. Specifically, one of the transitions enabled from q is selected randomly according to $\sigma(\alpha)$; then a target state is chosen according to the selected transition. The result of the action of σ is a pair $(q, \mu_{\sigma(\alpha)})$, which we call the *combined transition* according to $\sigma(\alpha)$, where $\mu_{\sigma(\alpha)}$ is a distribution of $SubDisc(\Sigma \times S)$ defined as follows: for each pair (a, s),

$$\mu_{\sigma(\alpha)}((a, s)) \triangleq \sum_{(q, a, \mu) \in \mathcal{D}} \sigma(\alpha)((q, a, \mu)) \mu(s).$$

The result of the action of a scheduler from a state s can be represented as a Markov process whose states are executions of $\mathcal{A}$ with start state s. We call these objects *probabilistic executions*. Formally, the probabilistic execution of $\mathcal{A}$ induced by a scheduler σ and a state s is a process (Q, μ) where Q is the set of executions of $\mathcal{A}$ that start with s, and μ is 0 except for pairs of the form (q, qar) where it is defined as follows: $\mu(q, qar) = \mu_{\sigma(q)}((a, r))$.

Given a scheduler σ and a state s we can define a probability measure $\mu_{\sigma,s}$ on the paths of the induced probabilistic execution. The measure is defined according to the cone construction, where a cone is a set of the form $C_\alpha = \{\alpha' \mid \alpha \le \alpha'\}$, α being a finite execution that starts with s, and $\mu_{\sigma,s}$ is defined on the σ-field generated by the cones. The definition of $\mu_{\sigma,s}$ on the cones is given inductively as follows:

1. $\mu_{\sigma,s}(C_s) \triangleq 1$;
2. $\mu_{\sigma,s}(C_{\alpha aq}) \triangleq \mu_{\sigma,s}(C_\alpha)\mu_{\sigma(\alpha)}((a, q))$.

It is a standard argument [4] to show that $\mu_{\sigma,s}$ is σ-additive and that it can be extended uniquely to the σ-field generated by the cones.

Throughout the paper we usually refer to a probabilistic execution as the corresponding measure over executions $\mu_{\sigma,s}$ and we omit either σ or s whenever they are not needed or clear from the context.

4   Coin Lemmas for a Single Experiment

In this section we give our new formulation of a coin lemma for a single experiment and we compare it with the formulation of [12]. We start with a simple example that illustrates the structure of a coin lemma.

Example 4. Consider the experiment of rolling a die with 6 faces. We know that each face has probability 1/6. If we have a probabilistic automaton where a transition encodes the action of rolling a die and where a beep signal is sent whenever the outcome of the die is an even number, then we could say that in each probabilistic execution the probability of observing a beep signal is 1/2. However, it may be the case that a die is not rolled in every probabilistic execution; thus, a correct statement would be that in each probabilistic execution the probability of either not rolling any die or observing a beep signal is at least 1/2. This simple observation is at the base of the formulation of a coin lemma. Although the statement seems to be obvious, when the number of experiments grows the result is not obvious any more, and indeed in several occasions the principle was not applied correctly.

4.1   The Old Formulation of a Coin Lemma

We formulate a coin lemma for a single experiment as it is formulated in [12].

Lemma 1. *Let $\mathcal{A}$ be a probabilistic automaton, and let (a, U) be a pair consisting of an action a of $\mathcal{A}$ and a set of states U of $\mathcal{A}$. Let p be a real number between 0 and 1 such that for each transition (s, a, μ) of $\mathcal{A}$, $\mu(U) \ge p$.*

For each probabilistic execution μ_σ of A let $FIRST(a, U)(\mu_\sigma)$ be the set of executions α of A such that either a does not occur in α, or a occurs in α and the state reached after the first occurrence of a is a state of U.

Then, for each probabilistic execution μ_σ of A, $\mu_\sigma(FIRST(a, U)(\mu_\sigma)) \geq p$.

The experiment described by the coin lemma above consists of performing a transition labeled by a, and checking whether a state from U is reached after performing a. If there are multiple occurrences of a-labeled transitions, then we observe the first occurrence. The hypothesis of the lemma requires that in each a-labeled transition the probability of reaching a state from U is at least p. The conclusion states that the probability of either not performing any a-labeled transition or reaching a state from U after performing the first a-labeled transition is at least p. It is easy to generalize Lemma 1 to a coin lemma where the i-th occurrence of an action a is observed.

Example 5. Returning to the example of rolling a die, we need a special action, say *roll*, to label all transitions where a die is rolled. The set U is the set of states where the die gives an even number, and p is $1/2$.

The rule of Lemma 1 that associates an event with each probabilistic execution considers all those executions where either a does not occur or the first occurrence of a is followed by a state from U. The coin lemma guarrantees that all the events returned by the rule have minimum probability p. If we show that some good property holds for each execution of the events returned by the rule, then we can conclude that the good property holds with probability at least p no matter how the nondeterminism is resolved. In particular we are forced to show that either a occurs always or that the good property holds whenever a does not occur.

4.2   The New Formulation of a Coin Lemma

The old formulation of coin lemmas relies strongly on the labels associated with the transitions of a probabilistic automaton. The rule to determine whether an experiment takes place (in the case of Lemma 1 the experiment takes place at the first occurrence of a) is fixed in the formulation of the coin lemma. Thus, for each kind of experiment a new coin lemma should be formulated. Furthermore, it is difficult to formulate coin lemmas based on non-trivial methods to determine when an experiment takes place (e.g., the first a after three consecutive a's lead to a state from U).

The fact that success means reaching a fixed set of states U gives us limited flexibility as well. An example of limited flexibility can be observed in the proof of correctness of the consensus algorithm of Ben Or [1, 11]. The experiment consists of flipping a coin and the successful result depends on the past history. Sometimes it is successful to obtain head and sometimes it is successful to obtain tail. A coin lemma formulated in terms of a unique set of states U does not suffice.

Rather than formulating ad-hoc coin lemmas each time, here we propose a new formulation of coin lemmas that relies on random variables both to identify the places where an experiment takes place and the successful results of an experiment. A coin lemma does not fix any more the rule to determine where an experiment takes place; rather, it takes the rule as an argument in the form of a random variable.

An experiment either takes place or does not take place. Thus, we can define a binary function that associates 0 to those executions where the experiment does not take place and 1 to those executions where the experiment takes place. We require the binary function to be measurable, thus a random variable. If an experiment takes place in a finite execution α, then the experiment takes place in any extension of α, since in any extension of α the prefix α has occurred. If an experiment takes place in α, then we can identify the exact point at which the experiment takes place by looking at the minimum prefix of α where the experiment takes place. The important concept for us is that an experiment must take place at some finite point. Thus, if an experiment takes place in an infinite execution α, then it must take place also at some finite prefix of α. The outcome of an experiment can be captured by some other random variable whose value is observed only after the experiment takes place.

We now start to capture formally the notion of an experiment, which consists of a random variable that identifies where the experiment takes place, called an experiment detector, and another random variable that describes the outcome of the experiment. We consider a generic probabilistic execution μ and we denote by $\mathcal{F}$ the σ-field on which μ is defined.

Definition 1. *Let X be a random variable on $\mathcal{F}$. We say that X is finitely determined if there exists a root function ρ_X such that, $\rho_X(\alpha)$ is a finite prefix of α for each execution α, and whenever $X(\alpha) \neq 0$ for some execution α, $X(\alpha') = X(\alpha)$ for each execution α' such that $\rho_X(\alpha) \leq \alpha' \leq \alpha$.*

We say that the random variable X is persistent if, whenever $X(\alpha) \neq 0$ for some execution α, that $X(\alpha') = X(\alpha)$ for each execution α' such that $\alpha \leq \alpha'$.

In other words, the random variable X is finitely determined if, whenever its value is not 0 in an infinite execution, we can identify a finite point after which its value is fixed; the random variable X is persistent if its value does not change once it is different from 0. The root function is not unique in general. Normally we can take $\rho_X(\alpha)$ to be the minimum prefix of α that satisfies the property of Definition 1. Although we do not state it explicitly in the rest of the paper, we always assume that a finitely determined random variable is equipped with its root function.

Definition 2. *A random variable is binary if it takes values in the set $\{0, 1\}$. An experiment detector is a finitely determined persistent binary random variable. An experiment is a pair (D, U) of persistent random variables where D is an experiment detector. An experiment is said to be finitely determined if U is finitely determined as well. We say that an experiment is a binary experiment if U is binary.*

Whenever the random variables D and U are binary, we can refer to D and U as sets and we can use the classical set notation when referring to D and U.

Observe that any measurable boolean condition on any random variable U gives rise to a binary variable. Thus, it is always possible to derive binary experiments from generic experiments.

Example 6. In the die example the experiment is a pair (D_d, U_d) where $D_d(\alpha) = 1$ whenever the action of rolling the die takes place in α, and $U_d(\alpha) = 1$ if the state reached after the die is rolled corresponds to an even outcome.

In the example of Lemma 1, the experiment is the pair (D_a, U_a) where $D_a(\alpha) = 1$ if α contains an occurrence of action a, while $U_a(\alpha) = 1$ if a occurs in α and the state reached in α after the first occurrence of a is a state of the set U.

In the formulation of Lemma 1 there is an hypothesis stating that the probability of reaching a state from U in a transition labeled by a is at least p. In other words, once the experiment takes place, the probability of success must be p. We capture this idea with the notion of a p-successful experiment, which is expressed in terms of conditional distributions. The root function is used to determine the point where the experiment takes place.

Definition 3. *Given a binary experiment (D, U) and a number $p \in [0, 1]$, we say that the experiment (D, U) is p-successful in a probabilistic execution μ if for each execution $\alpha \in D$ it is the case that either $\mu(C_{\rho(\alpha)}) = 0$ or $\mu(U | C_{\rho(\alpha)}) \geq p$.*

The first result about p-successful experiments is that the local condition imposed in the definition implies a global condition as well on the probability of success whenever the experiment takes place.

Lemma 2. *Let $\mathcal{A}$ be a probabilistic automaton, μ be a probabilistic execution of $\mathcal{A}$, and (D, U) a p-successful experiment for μ. Then, either $\mu(D) = 0$ or $\mu(U | D) \geq p$.*

Proof. If $\mu(D) = 0$ then we are done. Otherwise, let θ be the set of minimal roots given by ρ_D, i.e., the minimal elements in the image under ρ_D of the event $\{D = 1\}$. By definition of θ, $\mu(D) = \sum_{\alpha \in \theta} \mu(C_\alpha)$. Then, $\mu(U | D) = \mu(D \cap U)/\mu(D) = \sum_{\alpha \in \theta} \mu(C_\alpha)\mu(U | C_\alpha)/\sum_{\alpha \in \theta} \mu(C_\alpha)$. By hypothesis, since the elements of θ are roots, for each $\alpha \in \theta$ we have $\mu(U | C_\alpha) \geq p$. Thus, $\mu(U | D) \geq p$.

The global condition proved in Lemma 2 leads directly to our new formulation of a coin lemma for single binary experiments.

Proposition 1. *Let $\mathcal{A}$ be a probabilistic automaton, μ be a probabilistic execution, and (D, U) a p-successful experiment for μ. Then, $\mu(\neg D \vee U) \geq p$.*

Proof. $\mu(\neg D \vee U) = \mu(\neg D) + \mu(D \cap U) = \mu(\neg D) + \mu(D)\mu(U | D) \geq \mu(\neg D) + \mu(D)p \geq \mu(\neg D)p + \mu(D)p = p$.

Example 7. Lemma 1 is just a special case of Proposition 1. To show that the experiment (D_a, U_a) is p-successful we extend the cone notation by considering cones of executions that end with an action as well. Then, the root of an execution α that contains action a is α truncated at the first occurrence of a. The condition on the a-labelled transitions of Lemma 1 implies directly that (D_a, U_a) is p-successful. Observe that in any probabilistic execution μ the event $\neg D_a \vee U_a$ coincides with the event $FIRST(a, U)(\mu)$.

Using the structure of Proposition 1 we can formulate easily coin lemmas for the ith occurrence of an action a or more elaborate coin lemmas that identify the occurrence of any generic condition. An example is the coin lemma needed in [11] for the analysis of the randomized consensus algorithm of Ben Or [1]. In this case the experiment consists of flipping a coin and the successful outcome depends on what happened before flipping the coin, which can be represented easily with the random variable U.

5   Properties of Single Experiments

There are several properties of p-successful experiments that can be studied separately and that lead to new coin lemmas. In particular, we can derive generalizations of the coin lemmas of [12] that identify the first experiment that takes place among many.

Definition 4. *A sub-experiment of an experiment (D, U) is an experiment (D', U') such that $D' \subseteq D$, $\rho_{D'} = \rho_D$, and $U \subseteq U'$.*

In a sub-experiment we perform the actual experiment in fewer places and impose fewer restrictions for its success. The condition on the root function ensures that we are not moving the places at which an experiment takes place.

Proposition 2. *Any sub-experiment of a p-successful experiment is p-successful.*

Binary experiments can be combined, leading to new binary experiments.

Proposition 3. *Let $\mathcal{A}$ be a probabilistic automaton and μ be a probabilistic execution of $\mathcal{A}$. Let (C, U) be a p-successful experiment for μ and (D, V) be a q-successful experiment for μ. Let $E = C \cup D$ and $Z = (C \cap U) \cup (D \cap V)$. Then (E, Z) is a $min(p, q)$-successful experiment for μ.*

The experiment (E, Z) of Proposition 3 is considered to be successful whenever at least one of the two experiments (C, U) and (D, V) occurs and is successful. By hypothesis we know that when one experiment takes place, then it is successful with probability either p or q. Thus, no matter what experiment takes place, the probability of success is at least $min(p, q)$.

Example 8. We can combine sub-experiments and Proposition 3 to observe the result of the first experiment among two as follows. Consider two binary experiments (C, U) and (D, V). Let $C \leq D$ denote the detector of experiment C occurring not after experiment D (experiment D may not occur at all); similarly, let $D < C$ denote the detector of experiment D occurring before experiment C.

Observe that $(C \leq D, U)$ is a sub-experiment of (C, U) and $(D < C, V)$ is a sub-experiment of (D, V). By applying Proposition 3 to $(C \leq D, U)$ and $(D < C, V)$, we obtain an experiment that is successful whenever the first experiment that take place among C, D is successful.

Thus, the probability that either no experiment takes place or that the first experiment that takes place is successful is at least $min(p, q)$. By an inductive argument the same construction can be generalized to an arbitrary finite number of experiments.

Example 9. By keeping the setting of Example 8, if we use C to detect the first occurrence of an action a, D to detect the first occurrence of an action b, U to check whether a state from some set X is reached immediately after the occurrence of a, and V to check a state from some set Y is reached immediately after the occurrence of b, then, applying Proposition 3 to $(C \leq D, U)$ and $(D < C, V)$, together with Proposition 2, we derive the coin lemmas of [12] that deal with the outcome of the first action among two. Such coin lemmas are used in the analysis of the randomized dyning philosophers algorithm of Lehman and Rabin [6, 7].

One last result about binary experiments considers the union of two experiments with disjoint successful outcomes. In this case, the probabilities of success add up as expected.

Proposition 4. *Let $\mathcal{A}$ be a probabilistic automaton and μ be a probabilistic execution of $\mathcal{A}$. Let (D, U) be a p-successful experiment for μ and (D, V) be a q-successful experiment for μ such that $U \cap V = \emptyset$. Then $(D, U \cup V)$ is a p + q-successful experiment.*

6   Multiple Experiments

A stochastic process consists of a collection of experiments. In this section we describe several ways of combining experiments, leading to general formulation of coin lemmas. We first need some preliminary definitions to understand when two experiments can be treated as independent.

Definition 5. *We say that two experiments (C, U) and (D, V) are separated if for each execution $\alpha \in C \cap D$, $\rho_C(\alpha) \neq \rho_D(\alpha)$.*

We say that two finitely determined experiments (C, U) and (D, V) are ordered if they are separated and for each $\alpha \in C \cap D$ either

- *$\rho_C(\alpha) < \rho_D(\alpha)$ and $U(\alpha) \neq 0$ implies $\rho_U(\alpha) < \rho_D(\alpha)$, or*
- *$\rho_D(\alpha) < \rho_C(\alpha)$ and $V(\alpha) \neq 0$ implies $\rho_V(\alpha) < \rho_C(\alpha)$.*

Informally, two experiments are separated if they take place at different points in an execution, and two experiments are ordered if one experiment is completed (the outcome is observed) before the other experiment starts. Ordering of experiments is necessary to ensure that two experiments are independent.

6.1   Multiple Binary Experiments

Several times we are interested in observing the simultaneous success of a collection of experiments. The proposition below, which generalizes the coin lemmas about conjunction of [12], shows how to deal with two experiments. The generalization to any finite number of experiments follows by an inductive argument and can be derived from Proposition 6 in the next section.

Proposition 5. *Let A be a probabilistic automaton and μ be a probabilistic execution of A. Let (C, U) be a p-successful experiment for μ and (D, V) be a q-successful experiment for μ such that (C, U) and (D, V) are ordered. Then*
$$\mu((\neg C \cup U) \cap (\neg D \cup V)) \geq pq$$

In the coin lemma above we are interested in the outcome of two experiments. The ordering condition imposes that one experiment terminates before the other experiment starts. This condition avoids cases like those where the success of an experiment coincides with the failure of the other experiment, which would invalidate the lower bound on probabilities. In other words, since the experiments are performed at different points in time and their outcome is observed independently, the two experiments are guarranteed to be independent.

6.2   Multiple General Experiments

Our main objective is to derive a formulation of a coin lemma that is as general as possible and that does not depend too much on the underlying stochastic process as well as the underlying computational model. In this section we give two coin lemmas that are based on possibly countably many experiments whose outcomes are required to satisfy some specific property identified by a binary measurable function.

The first coin lemma that we propose deals with a scenario where the experiments may occur in any order within a probabilistic execution. To accept arbitrary orderings of experiments, each occurrence of an experiment is required to respect the same probability measure.

Proposition 6. *Let A be a probabilistic automaton and μ be a probabilistic execution of A. Let $\{(D_i, X_i)\}_{i \in I}$, be a finite or countable collection of pairwise ordered experiments with root functions $\{\rho_i\}_{i \in I}$. Suppose that for each $i \in I$ there exists a measure μ_i such that for each execution $\alpha \in D_i$, $\mu_i = \mu(\cdot | \rho_i(\alpha))$, and let ν be the measure $\times_{i \in I} \mu_i$, the product of the μ_i's as independent measures.*

Let X be $f(X_1, X_2, \cdots)$, where f is a binary measurable function defined on the X_i's. Let Y be a binary function such that $Y = 1$ iff there exists a collection of real numbers $\{y_i\}_{i \in I}$ such that, for each $i \in I$, either $y_i = X_i$ or $D_i = 0$, and such that $f(y_1, y_2, \cdots) = 1$.

Then, $\mu(Y) \geq \nu(X)$.

Proposition 6 considers a collection of experiments $\{(D_i, X_i)\}_{i \in I}$. These experiments are assumed to be pairwise ordered, so that there is no place where two distinct experiments take place at the same time. This ensures independence. Each time an experiment (D_i, X_i) takes place there is always the same distribution μ_i that describes X_i conditional on the specific occurrence of the experiment. The distribution ν describes the composition of the μ_i measures as independent measures.

Function f is a boolean test on the variables X_i's. Function Y checks whether function f has a chance to be successful given the outcome of the experiments that take place. In other words, if there is a possibility to fix arbitrarily the outcome of the experiments that do not take place so that f evaluates to 1, then Y evaluates to 1 as well. For this purpose, observe that the y_i's must coincide with the X_i's whenever the corresponding experiments take place.

In the coin lemma that follows the experiments are required to occur according to a predetermined order and no experiment may occur if the previous experiments have not occurred yet. In this last case we can simply look at the global outcome of an experiment without looking at each single occurrence. The main difference between the coin lemma below and Proposition 6 is in the conditions enforced on the measures μ_i's.

Proposition 7. *Let $\mathcal{A}$ be a probabilistic automaton and μ be a probabilistic execution of $\mathcal{A}$. Let $\{(D_i, X_i)\}_{i \in I}$, be a finite or countable collection of pairwise ordered experiments such that $D_i \leq D_j$ whenever $i > j$. For each $i \in I$ let μ_i be the measure $\mu(\cdot | D_i)$, i.e., the measure of X_i conditional on D_i. Let ν be the measure $\times_{i \in I} \mu_i$.*

Let X be $f(X_1, X_2, \cdots)$, where f is a binary measurable function defined on the X_i's. Let Y be a binary function such that $Y = 1$ iff there exists a collection of real numbers $\{y_i\}_{i \in I}$ such that, for each $i \in I$, either $y_i = X_i$ or $I_i = 0$, and such that $f(y_1, y_2, \cdots) = 1$.

Then, $\mu(Y) \geq \nu(X)$.

A simple consequence of Proposition 7 is that, for an experiment (D, X) and a real number q, $\mu(\neg I \vee (X \geq q)) \geq \mu(X \geq q | D)$. In particular, if $\mu(X \geq q | D) \geq p$, then $\mu(\neg D \vee (X \geq q)) \geq p$. The result is obtained by considering a function f that checks whether $X \geq q$.

Example 10. Consider a finite collection $(D_1, X_1), \cdots (D_k, X_k)$ of pairwise separated binary experiments, and suppose that each experiment (D_i, X_i) is p_i-successful for some probability p_i. Let f be the minimum of the X_i's. Then the random variable Y of Proposition 6 identifies all the executions of the set $(\neg D_1 \cup X_1) \cap \cdots \cap (\neg D_k \cup X_k)$. Furthermore, the lower bound given by Proposition 6 is $p_1 \cdots p_k$, thus proving the generalization of Proposition 5.

Example 11. If we consider a probabilistic automaton where a possibly infinite sequence of coin flips occurs and we identify each experiment detector D_i with the ith coin flip and X_i with the ith outcome, then Proposition 6 gives us the coin lemma for random walks that appears in [8].

7   Concluding Remarks

We have reformulated coin lemmas in terms of random variables and conditional distributions, thus generalizing all known coin lemmas and making them independent of the specific model of probabilistic automata. An important step is the formulation of the notion of p-successful experiments, which are an abstraction of the building blocks of the coin lemmas of [12]. The new formulation of coin lemmas highlights more precisely the fact that, in order for a system to function correctly with high probability, the adversarial scheduler should not be able to gain any advantage by avoiding to schedule any random draws.

The next step in the generalization of coin lemmas is to understand better what happens when we deal with expectations. So far the only result about expectations appears in [8], where an upper bound on the expected termination time for a random walk with barriers is studied within a framework with nondeterminism. The formulation of the coin lemma in [8] is specific to the problem under examination and it is not clear yet how to derive such bounds in general.

References

1. M. Ben-Or. Another advantage of free choice: completely asynchronous agreement protocols. In *Proceedings of the 2nd Annual ACM Symposium on Principles of Distributed Computing*, Montreal, Quebec, Canada, August 1983.
2. C. Derman. *Finite State Markovian Decision Processes.* Academic Press, 1970.
3. H. Hansson. *Time and Probability in Formal Design of Distributed Systems*, volume 1 of *Real-Time Safety Critical Systems.* Elsevier, 1994.
4. J.G. Kemeny, J.L. Snell, and A.W. Knapp. *Denumerable Markov Chains.* Graduate Texts in Mathematics. Springer-Verlag, second edition, 1976.
5. K.G. Larsen and A. Skou. Bisimulation through probabilistic testing. In *Conference Record of the 16th ACM Symposium on Principles of Programming Languages*, Austin, Texas, pages 344–352, 1989.
6. D. Lehmann and M. Rabin. On the advantage of free choice: a symmetric and fully distributed solution to the dining philosophers problem. In *Proceedings of the 8th Annual ACM Symposium on Principles of Programming Languages*, pages 133–138, January 1981.
7. A. Pogosyants and R. Segala. Formal verification of timed properties of randomized distributed algorithms. In *Proceedings of the 14th Annual ACM Symposium on Principles of Distributed Computing*, pages 174–183, August 1995.
8. A. Pogosyants, R. Segala, and N. Lynch. Verification of the randomized consensus algorithm of Aspnes and Herlihy: a case study. *Distributed Computing*, 13:155–186, July 2000.
9. M.O. Rabin. Probabilistic automata. *Information and Control*, 6:230–245, 1963.
10. I. Saias. Proving probabilistic correctness: the case of Rabin's algorithm for mutual exclusion. In *Proceedings of the 11th Annual ACM Symposium on Principles of Distributed Computing*, Quebec, Canada, August 1992.
11. R. Segala. *Modeling and Verification of Randomized Distributed Real-Time Systems.* PhD thesis, MIT, Dept. of Electrical Engineering and Computer Science, 1995. Also appears as technical report MIT/LCS/TR-676.
12. R. Segala. The essence of coin lemmas. In *Proceedings of PROBMIV'98*, volume 22 of *Electronic Notes in Theoretical Computer Science*, 1999.

MoDeST – A Modelling and Description Language for Stochastic Timed Systems

Pedro R. D'Argenio, Holger Hermanns, Joost-Pieter Katoen, and Ric Klaren

Formal Methods and Tools Group, Faculty of Computer Science
University of Twente, P.O. Box 217, 7500 AE Enschede, The Netherlands

Abstract. This paper presents a modelling language, called MoDeST, for describing the behaviour of discrete event systems. The language combines conventional programming constructs – such as iteration, alternatives, atomic statements, and exception handling – with means to describe complex systems in a compositional manner. In addition, MoDeST incorporates means to describe important phenomena such as non-determinism, probabilistic branching, and hard real-time as well as soft real-time (i.e., stochastic) aspects. The language is influenced by popular and user-friendly specification languages such as Promela, and deals with compositionality in a light-weight process-algebra style. Thus, MoDeST (*i*) covers a very broad spectrum of modelling concepts, (*ii*) possesses a rigid, process-algebra style semantics, and (*iii*) yet provides modern and flexible specification constructs.

1 Introduction

System design is primarily focussed on functional aspects. Non-functional aspects such as reliability and performance typically play a role – if at all – in the final stages of the design trajectory. To overcome this problem, sometimes identified as the insularity problem of performance engineering [17,14], it has been widely recognised that quantitative system aspects should be considered during the entire system design trajectory. Although a complete insight in the quantitative aspects might not be present at each design stage, even with partial information (or rough estimates) design alternatives may be rejected early due to unsatisfactory performance or dependability characteristics. For this purpose, modelling techniques used by system engineers or those that provide an easy migration path for users need to be adapted to take quantitative system aspects into account.

This has resulted in extensions of light-weight formal notations such as SDL and UML on the one hand, and the development of a whole range of more rigorous formalisms based on e.g., stochastic process algebras, or appropriate extensions of labelled transition systems (such as timed and probabilistic automata [1,31]). Light-weight notations are typically closer to engineering techniques, but lack a formal semantics; rigorous formalisms do have such formal semantics, but their learning curve is typically too steep from a practitioner's perspective. In this paper, we propose a description language that is intended

L. de Alfaro and S. Gilmore (Eds.): PAPM-PROBMIV 2001, LNCS 2165, pp. 87–104, 2001.

to have a rigid formal basis (i.e., semantics) and incorporates several ingredients from light-weight notations such as exception handling[1], modularisation, atomic statements, iteration, and simple data types. The semantics enables formal reasoning and provides a solid basis for the development of tool support whereas the light-weight ingredients are intended to pave the migration path towards engineers.

Important rationales behind the development of the description language, called MoDeST (Modeling and Description language for Stochastic Timed systems), are:

- *Orthogonality.* The language has been set up in an orthogonal way such that timing and probabilistic aspects can easily be added to (or omitted from) a specification if these aspects are of (no) relevance.
- *Usability.* Syntax and language constructs have been designed to be close to other commonly used languages. The syntax resembles that of the programming language C and the modelling language Promela [21]. Data modularisation concepts and exception handling mechanisms have been adopted from modern object-oriented programming languages such as Java [16]. Process algebraic constructs have been strongly influenced by FSP (Finite State Processes [24]) a simple, elegant calculus that is aimed at educational purposes.
- *Practical considerations.* The design of the language and the development of accompanying prototype tool-support have taken place hand-in-hand. Considerations about the tool handling of language constructs have been a driving force behind the language development.
- *Expressiveness.* We have identified a handful of semantic concepts which are well-established in the context of computer-aided verification and modelling formalisms for stochastic discrete event systems:
  (1) *Action nondeterminism* is often used in concurrent system design to leave parts of the description underspecified, and is an appropriate means to reflect that the order of events in concurrent executions is out of the control of a modeller.
  (2) *Probabilistic branching* is a way to include quantitative information about the likelihood of choice alternatives. This is especially useful to model randomized distributed algorithms, but also suitable to represent scheduling strategies, quantify data dependencies etc. on an abstract level.
  (3) *Clocks* are a means to represent real time and to specify the dynamics of a model in relation to a certain time or time interval, represented by a specific value of a clock.
  (4) *Delay nondeterminism* allows one to leave the precise timing of events unspecified. In many cases, the system dynamics depends on events taking place in some time interval (e.g., prior to a time-out) where it is left unspecified when in the interval the event will occur.

---

[1] Exception handling in specification languages has received scant attention. Notable exceptions are Enhanced-LOTOS [15] and Esterel [3].

(5) *Random variables* are often used to give quantitative information about the likelihood of a certain event to happen after or within a certain time interval.

While (1) and (2) affect the dynamics of a model via the (discrete) set of next events, (4) and (5) are means to affect the model dynamics by the (continuous) elapse of time. Thus, (1) and (4) describe two distinct types of nondeterminism, while (2) and (5) represent distinct types of probabilistic behaviour. We believe that each of these concepts is indispensable if striving for an integrated consideration of quantitative system aspects during the entire system design trajectory. However, we are not aware of any other formalism, model, or tool that is powerful enough to cover the complete spectrum spanned by this classification. Some approaches however come close, among them [29,4,10,7,26]. We achieve the full expressiveness by using a model that integrates timed automata [1](using the deadline style of [6]), stochastic automata [13,11], and (simple) probabilistic automata [31]. These three ingredient models have been selected from a wide range of possible alternative models. They were chosen because they complement each other very well and yield precisely the desired expressiveness. Due to their individual compositional properties, the resulting model is elegant to use in the context of a compositonal semantics for the language MoDeST.

We claim that the language eases the description of a wide range of systems, because, in summary, it combines a rigid formal semantics with the following key features:

- light-weight control structures such as iteration, and exception handling
- simple data types that can be user-defined using modularisation (packages)
- composition and abstraction mechanisms to structure specifications
- atomic statements to control the granularity of transitions
- nondeterministic and probabilistic alternatives
- nondeterministic and probabilistic timing

This paper presents the formal syntax and semantics of MoDeST and discusses the relationship to existing models for probabilistic systems. The reader interested in data and data type treatments in MoDeST is referred to [22].

Organisation of the paper. Section 2 introduces the language ingredients of MoDeST in an incremental way. Section 3 defines the syntax and semantics formally. Section 4 discusses the range of models covered by MoDeST. Section 5 briefly addresses some analysis techniques for MoDeST specifications. Finally, Section 6 concludes the paper. For the sake of clarity, this paper focuses on behavioural aspects of the semantics and omits considerations on data manipulation. A full version of this paper is available [23].

2   A Gentle Language Primer

This section introduces the core language features of MoDeST by specifying a real-time cashier. This is done in an incremental manner starting from an untimed, non-probabilistic description.

The system is informally described as follows. In a supermarket customers arrive at the cashing point and queue in order to pay their selected products. The customers provide their products on a conveyor belt and the cashier takes the products one-by-one from the belt (this is modelled by action *get_prod*). The prod-

```
process Cashier() {
    do{:: get_prod ; alt {
                :: cash
                :: set_price ;
                    cash }
        }
    }
```

uct is either cashed (action *cash*), or in case there is no price tag, the cashier calls for assistance to establish the price (action *set_price*) after which cashing takes place (action *cash*). This behaviour is described by the above process, where ; denotes sequential execution and :: is used as a separator for the different alternatives of the choice construct **alt**. This construct is a way to model action nondeterminism. The cashier repeats his (or her) behaviour (indicated by **do**{:: ...} which is executed repeatedly, unless a **break** occurs).

In case more information is available about the likelihood with which a customer delivers a product without price tag, the nondeterministic choice may be replaced by a probabilistic choice. This yields the process depicted on the right, where weights (in the form of positive reals) are used to determine the likelihood

```
process Cashier() {
    do{:: get_prod palt {
                :49: cash
                : 1: set_price;
                    cash }
        }
    }
```

with which a certain alternative should be chosen. Here, price information is available with probability 0.98 and the price tag is absent with probability 0.02. In the terminology of Section 1, **palt** is a means to incorporate probabilistic branching. Each probabilistic choice-construct is required to be action guarded, i.e., immediately preceded by an action.

Another uncommon but very serviceable language construct is the possibility to raise and handle exceptions. To illustrate this concept, we slightly adapt the description of the cashier as depicted on the right. In case a product cannot be cashed due to an absent price tag, the cashier calls for assistance by raising an exception (modelled by action *no_price* of exception type). On handling this

```
process Cashier() {
    do{:: try { get_prod palt {
                :49: cash
                : 1: throw(no_price) }
        }
        catch no_price {
                set_price;
                cash }
        }
    }
```

exception the price is determined and the product is cashed.

In a construct like **try** { P } **catch** e { Q } the body P in general models the normal behaviour, whereas if action e occurs while executing P, an exception is raised that shall be handled by Q, i.e., control is passed from P to Q. Note that compared to our previous specification, an additional action (of exception type) has been introduced to signal the occurrence of the exceptional situation.

So far, our descriptions were time-less, i.e., we did not include any timing considerations with respect to the activities involved. In the next step, we will put some simple timing constraints on the cashier. Like in timed automata [1], the elapse of time in MoDeST is modelled by means of clock variables. Values of clock variables increase linearly as time progresses. For instance, in order to impose a delay of at least 120 time units between catching the exception *no_price* and determining the price of the product at hand (*set_price*),

```
process Cashier() {
    do{:: try { get_prod palt {
            :49: cash
            : 1: throw(no_price) }
        }
        catch no_price {
            y = 0;
            when(y ≥ 120)
                set_price;
            cash }
    }
}
```

we equip the previous description with clock variable y, and obtain the process on the right. Clock y is reset just after catching the exception *no_price* and the price can be determined at any time point after a delay of at least 120 time-units as indicated by the when-clause. In fact, each action needs to be preceded by a **when()** constraint, but unless otherwise specified **when(true)** is a default constraint (that can be omitted).

When-clauses thus indicate when a certain action may (i.e. is allowed to) happen. Similar to location invariants in safety timed automata [18] and deadlines in timed automata with deadlines [6], we need a separate mechanism to *force* certain actions to happen at some time instant. To that end, we use deadlines. For instance, the process on the right specifies that *set_price* is enabled from 120 time units after catching the exception (as before), and that it should happen before 240 time units after the catch

```
process Cashier() {
    do { try { get_prod palt {
            :49: cash
            : 1: throw(no_price) }
        }
        catch no_price {
            y = 0;
            urgent(y ≥ 240)
            when(y ≥ 120)
                set_price;
            cash }
    }
}
```

– as indicated by the urgent-clause. More precisely, if the exception is catched at time t, say, then *set_price* will happen at some time instant $t+\Delta$ where Δ is nondeterministically chosen from the closed interval $[120,240]$. Thus, differences in guards and deadline constraints induce delay nondeterminism.

In general, if an action is guarded by **urgent(B)**, for boolean expression B, it must be executed as soon as B becomes true. Therefore, a system is allowed to idle as long as none of its activities becomes urgent. The language user can influence whether by convention activities are assumed to be urgent (guarded by **urgent(true)**), or non-urgent (guarded by **urgent(false)**), via setting a flag in the preamble of a MoDeST specification.

As a next step, we impose a delay on the cashing of the cashier, i.e., on action *cash*. Depending on (the price of) the product, environmental circumstances (such as the mood of the cashier, the time of the day), and so on, the duration of cashing may vary. We assume that cashing takes between 10 and 20 time-units. If no more information is available this could be modelled in a similar way as we just treated *set_price*. However, we now assume that the duration of cashing is uniformly distributed over the interval $[10, 20]$. In this case, the modelling as just above does not suffice, as it would choose a time instant nondeterministically without taking the likelihoods into account. To that end, we equip the specification with a clock variable x, say, and add a float variable xd, say,

```
process Cashier() {
    do {:: try { get_prod palt {
            :49: Cashing()
            : 1: urgent(true)
                throw(no_price) }
        }
        catch no_price {
            y = 0;
            urgent(y ≥ 240)
            when(y ≥ 120)
                set_price;
            Cashing() }
    }
}

process Cashing() {
    [xd = U[10, 20], x = 0];
    urgent(x ≥ xd)
    when(x ≥ xd)
        cash
}
```

that is used to store a sample value drawn from a probability distribution. Thus, the occurrences of *cash* in process *Cashier* is replaced by invoking a process *Cashing* depicted on the right. In the latter, the statement [...] contains a set of assignments that are executed atomically, i.e., without interference with executions of other processes in the system. In this example, the variable xd is assigned a (float) value according to a uniform distribution on interval $[10, 20]$, and clock x is reset. The urgent- and when-clause make sure that *cash* takes place as soon as x has reached the value xd.

The overall system could be modelled by, for instance, the expression on the right, where N is the parameter (i.e., the length) of the queue. Variables do not need to be declared globally, a variable (or action, or exception) can equally well be declared local to a process. Processes are put in parallel via the **par**{::...} construct. These processes execute their activities independently from each other, except that common (non-local) actions

```
exception no_price;
clock x, y;
float xd;
patient get_prod, cash, set_price;

par {
    :: Arrivals() ;
    :: Queue(N) ;
    :: Cashier()
}
```

need to be executed synchronously, à la CSP [20]. One of the keywords appearing in the preamble needs further explanation. We distinguish **patient** and **impatient** actions. If a patient action is common to multiple processes, then the synchronized action becomes urgent as soon as *all* partners require urgency. In contrast, a process that intends to synchronise on an impatient action is not willing to wait for the partner. Thus a synchronized impatient action is urgent as soon as *at least one* synchronization partner requires urgency.

3 Formal Definition of MoDeST

This section formally defines the language MoDeST, the underlying operational model, and the operational semantics of MoDeST. The semantics maps each MoDeST specification on some *stochastic timed automata* (STA, for short). STA combine the power of timed automata [1] using the deadline style of [6], stochastic automata [13,11], and (simple) probabilistic automata [31]. Before discussing syntax and semantics of MoDeST we introduce STA together with other relevant concepts.

3.1 Stochastic Timed Automata

A *probability space* is a tuple $(\Omega, \mathcal{F}, \mathbf{P})$ where Ω is the *sample space*, $\mathcal{F}$ is a σ-*algebra* containing subsets of Ω, and $\mathbf{P}$ is a *probability measure* on the measurable space $(\Omega, \mathcal{F})$. If $\mathcal{P}$ is a probability space, we write $\Omega_{\mathcal{P}}$, $\mathcal{F}_{\mathcal{P}}$ and $\mathbf{P}_{\mathcal{P}}$ for its sample space, σ-algebra, and probability measure, respectively[2]. Let $Prob(H)$ denote the set of probability spaces $(\Omega, \mathcal{F}, \mathbf{P})$ such that $\Omega \subseteq H$.

Let Var be a set of typed *variables* with a distinguished subset $\mathsf{Ck} \subseteq \mathsf{Var}$ of *clock variables* (variables of type **clock**). Let RVar be a (finite) set of *random variables* such that $\mathsf{RVar} \cap \mathsf{Var} = \emptyset$. Let Exp be a set of *expressions* with variables in $\mathsf{Var} \cup \mathsf{RVar}$. Let $\mathsf{BExp} \subseteq \mathsf{Exp}$ be the set of *boolean expressions*, ranged over by $d, d', g, g' \dots$. A boolean expression is required not to contain random variables. $A : \mathsf{Var} \to \mathsf{Exp}$, is called an *assignment*. Let Assign denote the set of assignments. Let Act be a set of *action names*. We use a to range over elements of Act.

Definition 1. *A stochastic timed automaton (STA) is a triple* $(\mathcal{S}, \mathsf{Act}, \to)$, *where* $\mathcal{S}$ *is a set of* locations *and* $\to \, \subseteq \, \mathcal{S} \times \mathsf{Act} \times \mathsf{BExp} \times \mathsf{BExp} \times Prob(\mathsf{Assign} \times \mathcal{S})$.

For $\langle s, a, g, d, \mathcal{P} \rangle \in \to$, we write $s \xrightarrow{a,g,d} \mathcal{P}$ and require that $\mathcal{P}$ is a *discrete* probability space. We call g the *guard* and d the *deadline*. Intuitively, the system is allowed to execute an edge $s \xrightarrow{a,g,d} \mathcal{P}$ whenever it is at location s and the guard g holds under the current values of the variables. If in addition the deadline d holds, then the system is obliged to execute the edge before time progresses. Due to this fact, the system is allowed to wait in location s as long as no deadline in one of its outgoing edges becomes true. Once the edge $s \xrightarrow{a,g,d} \mathcal{P}$ is executed, the system moves to location s' assigning values according to A with probability $\mathbf{P}_{\mathcal{P}}(\langle s', A \rangle)$.

---

[2] We assume familiarity with the basics of probability and measure theory (see e.g. [32]).

Depicted on the right is an example STA corresponding to the final *Cashier* specification of Section 2. Locations are represented by circles. A probabilistic edge is represented by a solid line from which doted arrows fan out. The solid line is labelled by the guard, deadline, and synchronisation label. Each dotted arrow represents a probabilistic alternative, and are labelled with a probability value and a set of assignments. Their target is the next location. Dead-

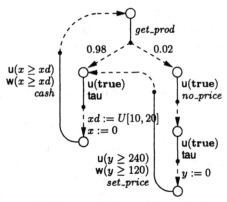

lines are prefixed by a 'u' (urgent) and omitted if they are **false**, and guards by a 'w' (when) and omitted whenever they are **true**. Trivial probabilities and empty assignments are also omitted.

STA provide a symbolic framework to represent stochastic timed behaviour, but this representation is too abstract to represent the concrete evolution as describe above, which is needed for different kinds of analysis, such as probabilistic model checking, or discrete event simulation. Therefore, STA have an interpretation in terms of timed continuous probabilistic transition systems. This interpretation is given in [23].

3.2   Syntax

In the following we discuss the language constructs of MoDeST. We assume that the set of actions Act consists disjointly of:

- a set PAct of *patient actions,*
- a set IAct of *impatient actions,*
- a set Excep of *exception names,*
- an action $\perp$ indicating an *unhandled error,*
- an action **break** indicating the *breaking of a loop,* and
- an action **tau** indicating an unobservable activity called *silent step.*

The set of processes of the language MoDeST is given by the following grammar,

$$
\begin{aligned}
P ::= \quad &\textbf{stop} &&| \; \textbf{error} &&| \; ProcName(e_1, \ldots, e_k)\\
&| \; \textbf{when}(b) \; P &&| \; \textbf{urgent}(b) \; P &&| \; \textbf{alt}\{::P_1 \ldots ::P_k\}\\
&| \; act &&| \; act \; \textbf{palt} \; \{:w_1:asgn_1; \; P_1 \ldots :w_k:asgn_k; \; P_k\}\\
&| \; \textbf{throw}(excp) &&| \; \textbf{try}\{P\} \; \textbf{catch} \; excp_1 \; \{P_1\} \ldots \textbf{catch} \; excp_k \; \{P_k\}\\
&| \; \textbf{break} &&| \; \textbf{do}\{::P_1 \ldots ::P_k\}\\
&| \; P_1; P_2 &&| \; \textbf{par}\{::P_1 \ldots ::P_k\}\\
&| \; \textbf{hide}\{act_1, \ldots, act_k\} \; P \quad | \quad \textbf{extend}\{act_1, \ldots, act_k\} \; P\\
&| \; \textbf{relabel} \; \{act_1, \ldots, act_k\} \; \textbf{by} \; \{act'_1, \ldots, act'_k\} \; P
\end{aligned}
$$

where, for $1 \leq i \leq k$, w_i is a positive integer representing a *weight*, $act, act'_i \in$ PAct $\cup$ IAct $\cup$ {tau}, $act_i \in$ PAct $\cup$ IAct, $excp, excp_i \in$ Excep, $b \in$ BExp, $e_i \in$ Exp not containing random variables, and $asgn_i$ is a list of assignments of the form $[x_1 = e_1, x_2 = e_2, \ldots, x_n = e_n]$. A MoDeST process is defined by

$$\textbf{process } ProcName(t_1 \ x_1, \ldots, t_k \ x_k) \ \{dcl \ P\}$$

where $\{x_1, \ldots, x_k\} \in$ Var, $\{t_1, \ldots, t_k\}$ are valid types, dcl is a sequence of declarations possibly including process definitions, $ProcName$ is a process name and P is as before. We write **process** $ProcName(x_1, \ldots, x_k)$ $\{P\}$ instead in the remainder of this paper, for convenience.

Each set $[x_1 = e_1, \ldots, x_n = e_n]$ induces a unique assignment $A \in$ Assign defined by $A(x_i) = e_i$, for $1 \leq i \leq n$, and $A(y) = y$ if $y \notin \{x_1, \ldots, x_n\}$. Therefore, we use $[x_1 = e_1, \ldots, x_n = e_n] \in$ Assign to refer to its induced assignment $A \in$ Assign.

MoDeST provides some further useful operations which are shorthand notations for some common constructions. They are described in Appendix A.

3.3   Semantics

The operational semantics of MoDeST is defined in terms of the stochastic timed automaton $(\mathcal{S}, \text{Act}, \twoheadrightarrow)$ where the set of locations $\mathcal{S}$ is defined by the set of MoDeST processes extended with a special termination mark $\sqrt{}$. The relation $\twoheadrightarrow$ is defined in the remainder of this section. In the following we use $\mathbf{Trv}(r)$ to denote the *trivial probability space* with sample space $\{r\}$. We also resort to *measurable functions*. Recall that $\mathbf{M} : \Omega_1 \to \Omega_2$ is measurable if $\mathbf{M}^{-1}(C) \in \mathcal{F}_1$ for all $C \in \mathcal{F}_2$ and that it induces a probability space $\mathbf{M}(\Omega_1, \mathcal{F}_1, \mathbf{P}_1) = (\Omega_2, \mathcal{F}_2, \mathbf{P}_1 \circ \mathbf{M}^{-1})$. In our case, all measurable functions are defined to be surjective. Under this condition $\Omega_2 = \mathbf{M}(\Omega_1)$.

Primitive operators. **stop** does not perform any activity and as such it does not produce any transition. act performs action act with no restriction and then terminates. **break**, used to break a **do** loop, can perform action **break** with no restriction and then terminates. **error** is a process that indicates an unhandled error by persistent executions of action $\perp$. The last of the basic operations, **throw**($excp$), raises an exception by executing action $excp \in$ Excep. If it is not handled, the system ends up in an unhandled error. In all these cases, urgency of the execution depends on a global boolean variable **urge** which can be set to **true** or **false** in the preamble section of the specification. If set to **true**, the specified system responds to *maximal progress* (default is **false**). We get:

$$act \xrightarrow{\ act, \textbf{true}, \textbf{urge}\ } \mathbf{Trv}(\sqrt{}) \qquad\qquad error \xrightarrow{\ \perp, \textbf{true}, \textbf{urge}\ } \mathbf{Trv}(error)$$

$$break \xrightarrow{\ break, \textbf{true}, \textbf{urge}\ } \mathbf{Trv}(\sqrt{}) \qquad\qquad throw(excp) \xrightarrow{\ excp, \textbf{true}, \textbf{urge}\ } \mathbf{Trv}(error)$$

Probabilistic prefix. act palt $\{:w_1:asgn_1; P_1 \ldots :w_k:asgn_k; P_k\}$ performs action *act* with no restriction, but as urgently as indicated by urge. Simultaneously, it randomly selects an alternative $i \in \{1, \ldots, k\}$ according to the weights $w_1, \ldots, w_k$, performs an assignment according to $asgn_i$, and continues executing P_i.

$$act\ \text{palt}\ \{:w_1:asgn_1; P_1 \ldots :w_k:asgn_k; P_k\} \xrightarrow{\ act,\text{true},\text{urge}\ } \mathcal{P}$$

where $\mathcal{P}$ is a discrete probability space with $\Omega_\mathcal{P} = \{\langle asgn_i, P_i \rangle \mid 1 \leq i \leq k\}$ and

$$\mathbf{P}_\mathcal{P}(\langle asgn_i, P_i \rangle) \overset{\text{def}}{=} \frac{w_i \cdot \#\{j \mid 1 \leq j \leq k \wedge asgn_i = asgn_j, P_i = P_j\}}{\sum_{j=1}^k w_j}$$

Conditions. when(b) P restricts the next activity of P to be performed whenever b holds. urgent(b) P enforces P to be urgent whenever b holds:

$$\frac{P \xrightarrow{a,g,d} \mathcal{P}}{\text{when}(b)\ P \xrightarrow{a,b \wedge g,d} \mathcal{P}} \qquad\qquad \frac{P \xrightarrow{a,g,d} \mathcal{P}}{\text{urgent}(b)\ P \xrightarrow{a,g,b \vee d} \mathcal{P}}$$

Choice. alt$\{::P_1 \ldots ::P_k\}$ executes precisely one P_i, selected in a nondeterministic fashion:

$$\frac{P_i \xrightarrow{a,g,d} \mathcal{P}_i \qquad (1 \leq i \leq k)}{\text{alt}\{::P_1 \ldots ::P_k\} \xrightarrow{a,g,d} \mathcal{P}_i}$$

Loop. do$\{::P_1 \ldots ::P_k\}$ repeatedly chooses a nondeterministic alternative. The execution finishes when one of the processes executes a break. We define the semantics of do in terms of alt and an auxiliary operator auxdo:

$$\text{do}\{::P_1 \ldots ::P_k\} \overset{\text{def}}{=} \text{auxdo}\{\text{alt}\{::P_1 \ldots ::P_k\}\}\{\text{alt}\{::P_1 \ldots ::P_k\}\}$$

The semantics of auxdo is given by:

$$\frac{P \xrightarrow{a,g,d} \mathcal{P} \qquad (a \neq \text{break})}{\text{auxdo}\{P\}\{Q\} \xrightarrow{a,g,d} \mathbf{M}_{\text{do}}(\mathcal{P})} \qquad\qquad \frac{P \xrightarrow{\text{break},g,d} \mathcal{P}}{\text{auxdo}\{P\}\{Q\} \xrightarrow{\text{tau},g,d} \mathcal{P}}$$

where $\mathbf{M}_{\text{do}}(\langle A, P' \rangle) \overset{\text{def}}{=} \langle A, \text{auxdo}\{P'\}\{Q\} \rangle$, if $P' \neq \sqrt{}$, and otherwise, $\mathbf{M}_{\text{do}}(\langle A, \sqrt{} \rangle) \overset{\text{def}}{=} \langle A, \text{auxdo}\{Q\}\{Q\} \rangle$.

Exception handling. The process try$\{P\}$ catch $excp_1$ $\{P_1\}$ … catch $excp_k$ $\{P_k\}$ executes P and terminates if P terminates without raising any exception beforehand. If instead P raises an exception $excp_i$, it is handled by executing the respective process P_i:

$$\frac{P \xrightarrow{a,g,d} \mathcal{P} \qquad (a \notin \{excp_1, \ldots, excp_k\})}{\text{try}\{P\}\ \text{catch}\ excp_1\ \{P_1\} \ldots \text{catch}\ excp_k\ \{P_k\} \xrightarrow{a,g,d} \mathbf{M}_{\text{try}}(\mathcal{P})}$$

$$\frac{P \xrightarrow{excp_i,g,d} \mathcal{P} \qquad (1 \leq i \leq k)}{\text{try}\{P\}\ \text{catch}\ excp_1\ \{P_1\} \ldots \text{catch}\ excp_k\ \{P_k\} \xrightarrow{\text{tau},g,d} \mathbf{Trv}(P_i)}$$

Table 1. Alphabet of a MoDeST term

$\alpha(\mathsf{stop}) = \alpha(\mathsf{error}) = \alpha(\mathsf{break}) = \alpha(\mathsf{throw}(excp)) = \emptyset$

$\alpha(act) = \{act\} - \{\mathsf{tau}\}$

$\alpha(act \ \mathsf{palt} \ \{:w_1{:}asgn_1;\ P_1 \ \ldots \ :w_k{:}asgn_k;\ P_k\}) = \alpha(act) \cup \bigcup_{i=1}^{k} \alpha(P_i)$

$\alpha(\mathsf{when}(b)\ P) = \alpha(\mathsf{urgent}(b)\ P) = \alpha(P)$

$\alpha(\mathsf{alt}\{::P_1 \ \ldots \ ::P_k\}) = \alpha(\mathsf{do}\{::P_1 \ \ldots \ ::P_k\}) = \alpha(\mathsf{par}\{::P_1 \ \ldots \ ::P_k\}) = \bigcup_{i=1}^{k} \alpha(P_i)$

$\alpha(P_1;\ P_2) = \alpha(P_1) \cup \alpha(P_2)$

$\alpha(\mathsf{try}\{P\}\ \mathsf{catch}\ excp_1\ \{P_1\} \ \ldots \ \mathsf{catch}\ excp_k\ \{P_k\}) = \alpha(P) \cup \bigcup_{i=1}^{k} \alpha(P_i)$

$\alpha(\mathsf{hide}\{act_1,\ldots, act_k\}\ P) = \alpha(P) - \{act_1,\ldots, act_k\}$

$\alpha(\mathsf{relabel}\ \{act_1,\ldots, act_k\}\ \mathsf{by}\ \{act'_1,\ldots, act'_k\}\ P) =$
$$\alpha(P)[act_1/act'_1,\ldots, act_k/act'_k] - \{\mathsf{tau}\}$$

$\alpha(\mathsf{extend}\{act_1,\ldots, act_k\}\ P) = \alpha(P) \cup \{act_1,\ldots, act_k\}$

$\alpha(ProcName(e_1,\ldots, e_k)) = \alpha(P) \qquad \text{provided } \mathsf{process}\ ProcName(x_1,\ldots, x_k)\ \{P\}$

---

where $\mathbf{M}_{\mathsf{try}}(\langle A, P'\rangle) \stackrel{\text{def}}{=} \langle A, \mathsf{try}\{P'\}\ \mathsf{catch}\ excp_1\ \{P_1\} \ \ldots \ \mathsf{catch}\ excp_k\ \{P_k\}\rangle$, if $P' \neq \surd$, and $\mathbf{M}_{\mathsf{try}}(\langle A, \surd\rangle) \stackrel{\text{def}}{=} \langle A, \surd\rangle$.

Sequential composition. $P_1;\ P_2$ executes P_1 until it finishes. Then it continues with the execution of P_2:

$$\frac{P_1 \xrightarrow{a,g,d} \mathcal{P}}{P_1;\ P_2 \xrightarrow{a,g,d} \mathbf{M}_;(\mathcal{P})}$$

where $\mathbf{M}_;(\langle A, P'\rangle) \stackrel{\text{def}}{=} \langle A, P';\ P_2\rangle$, if $P' \neq \surd$, and $\mathbf{M}_;(\langle A, \surd\rangle) \stackrel{\text{def}}{=} \langle A, P_2\rangle$.

Parallel composition. $\mathsf{par}\{::P_1 \ \ldots \ ::P_k\}$ executes processes $P_1,\ldots, P_k$ concurrently, synchronising them on the intersected alphabet, therefore allowing multiway synchronisation. The alphabet of a process P is the set $\alpha(P) \subseteq \mathsf{PAct} \cup \mathsf{IAct}$ of all actions P recognises. It is formally defined in Table 1. To define the semantics of MoDeST parallel composition, we resort to the auxiliary operator $\|_B$, with $B \subseteq \mathsf{PAct} \cup \mathsf{IAct}$, that behaves like CSP or LOTOS parallel composition [20,5]. Thus, par is defined by

$$\mathsf{par}\{::P_1 \ \ldots \ ::P_k\} \stackrel{\text{def}}{=} (\ldots((P_1 \|_{B_1} P_2) \|_{B_2} P_3)\ldots) \|_{B_{k-1}} P_k$$

with $B_j = (\bigcup_{i=1}^{j} \alpha(P_i)) \cap \alpha(P_{j+1})$. The behaviour of $\|_B$ is formally defined by the following rules (we omit the symmetric rule of interleaving):

$$\frac{P_1 \xrightarrow{a,g,d} \mathcal{P} \quad (a \notin B)}{P_1 \|_B P_2 \xrightarrow{a,g,d} \mathbf{M}_{\mathsf{par}P_2}(\mathcal{P})} \qquad \frac{P_1 \xrightarrow{a,g_1,d_1} \mathcal{P}_1 \quad P_2 \xrightarrow{a,g_2,d_2} \mathcal{P}_2 \quad (a \in B)}{P_1 \|_B P_2 \xrightarrow{a,g_1 \wedge g_2, d_1 \lozenge d_2} \mathbf{M}_{\mathsf{par}}(\mathcal{P}_1 \times \mathcal{P}_2)}$$

where $d_1 \Diamond d_2 = d_1 \wedge d_2$ if $a \in \mathsf{PAct}$ (that is, if the synchronising action is patient) and $d_1 \Diamond d_2 = d_1 \vee d_2$ otherwise (impatient). The operator $\times$ denotes the usual product on probabilistic spaces, and $\mathbf{M}_{\mathsf{par}P_2}(\langle A, P' \rangle) \stackrel{\text{def}}{=} \langle A, P' \|_B P_2 \rangle$, if $P' \neq \surd$ or $P_2 \neq \surd$, otherwise $\mathbf{M}_{\mathsf{par}\surd}(\langle A, \surd \rangle) \stackrel{\text{def}}{=} \langle A, \surd \rangle$. Furthermore,

$$\mathbf{M}_{\mathsf{par}}(\langle A_1, P_1' \rangle, \langle A_2, P_2' \rangle) \stackrel{\text{def}}{=} \begin{cases} \text{if } A_1 \cup A_2 \text{ is not a function then} \\ \qquad \langle \emptyset, \text{throw } inconsistency \rangle \\ \text{else} \\ \qquad \langle A_1 \cup A_2, P_1' \|_B P_2' \rangle \quad \text{if } P_1' \neq \surd \text{ or } P_2' \neq \surd \\ \qquad \langle A_1 \cup A_2, \surd \rangle \qquad \text{if } P_1' = P_2' = \surd \end{cases}$$

Some remarks are in order. A parallel composition terminates whenever all its components terminate. Moreover, notice that the difference between synchronisation of patient and impatience actions is only given by the way the deadlines are related. Since a process that wants to synchronise on a patient action always waits for its partner to be ready, then its deadline needs to be relaxed to the requirements of the partner. As a consequence, a deadline in a patient synchronisation is met whenever all the components meet their respective deadlines. Instead, a process that intends to synchronise on an impatient action is not willing to wait for the partner. Therefore, a deadline in an impatient synchronisation should be met as soon as one of the one of the synchronising components meets its deadlines. Finally, remark that during synchronisation an inconsistency of assignments may arise due to different write accesses to the same variable, i.e., if $A_1(x) \neq A_2(x)$ for some variable x. We treat this situation by raising a predefined exception.

Relabelling and hiding. relabel $\{act_1, \ldots, act_k\}$ by $\{act_1', \ldots, act_k'\}$ P behaves like P except that every action act_i is renamed by the corresponding act_i':

$$\frac{P \xrightarrow{a,g,d} \mathcal{P} \qquad f = [act_1/act_1', \ldots, act_k/act_k']}{\mathsf{relabel} \ \{act_1, \ldots, act_k\} \text{ by } \{act_1', \ldots, act_k'\} \ P \xrightarrow{f(a),g,d} \mathbf{M}_{\mathsf{relabel}}(\mathcal{P})}$$

where $\mathbf{M}_{\mathsf{relabel}}(\langle A, P' \rangle) \stackrel{\text{def}}{=} \langle A, \mathsf{relabel} \ \{act_1, \ldots, act_k\}$ by $\{act_1', \ldots, act_k'\} \ P' \rangle$, if $P' \neq \surd$, otherwise $\mathbf{M}_{\mathsf{relabel}}(\langle A, \surd \rangle) \stackrel{\text{def}}{=} \langle A, \surd \rangle$.

Hiding is a particular form of relabeling in which actions are renamed by the silent action tau. Therefore we define:

$$\mathsf{hide}\{act_1, \ldots, act_k\} \ P \stackrel{\text{def}}{=} \mathsf{relabel} \ \{act_1, \ldots, act_k\} \text{ by } \underbrace{\{\mathsf{tau}, \ldots, \mathsf{tau}\}}_{k \text{ times}} \ P$$

Alphabet extension. extend is only used to extend the alphabet that a process recognises (see Table 1). Otherwise, it does not affect the behaviour:

$$\frac{P \xrightarrow{a,g,d} \mathcal{P}}{\mathsf{extend}\{act_1, \ldots, act_k\} \ P \xrightarrow{a,g,d} \mathbf{M}_{\mathsf{extend}}(\mathcal{P})}$$

where $\mathbf{M_{extend}}(\langle A, P' \rangle) \stackrel{\text{def}}{=} \langle A, \mathbf{extend}\{act_1, \dots, act_k\} P' \rangle$,   and
$\mathbf{M_{extend}}(\langle A, \sqrt{} \rangle) \stackrel{\text{def}}{=} \langle A, \sqrt{} \rangle$.

Process instantiation. Provided **process** $ProcName(x_1, \dots, x_k)$ $\{P\}$ is part of the MoDeST specification under consideration, $ProcName(e_1, \dots, e_k)$ behaves like P where variables $x_1, \dots, x_k$ are substituted by their respective instantiations $e_1, \dots, e_k$.

$$\frac{P[x_1/e_1, \dots, x_k/e_k] \xrightarrow{a,g,d} \mathcal{P}}{ProcName(e_1, \dots, e_k) \xrightarrow{a,g,d} \mathcal{P}}$$ provided **process** $ProcName(x_1, \dots, x_k)\{P\}$

In summary, the relation $\rightarrow$ is the least relation satisfying the above rules. The reader is invited to check that the STA depicted in Section 3.1 is derived from the final *Cashier* specification of Section 2 using these semantic rules (see Appendix A for the shorthand notations used).

4   Derivable Models

MoDeST is expressive enough to cover a wide range of timed, probabilistic, non-deterministic, and stochastic models. These submodels play a crucial role in the context of analysing MoDeST specifications. Table 2 lists a range of prominent models and makes precise which semantic concepts (cf. Section 1) each of them shares with STA.

LTS: Labelled transition systems are the basic models of concurrency, they are usually analysed with techniques such as model checking or equivalence checking. They arise from MoDeST by disallowing the use of all time and stochastic concepts.

PTS: Probabilistic transition systems are labelled transition systems where some state changes are governed by discrete probability distributions while others are nondeterministic. They can be analysed with techniques from Markov decision theory, model checking, and equivalence checking [31,9]. MoDeST subsumes (simple) PTS via the **palt** construct which is action guarded by default.

Table 2. Submodels of STA

	LTS	PTS	TA	PTA	MC	GSMP	IMC	SA	STA
probabilistic branching	NO	YES	NO	YES	YES	YES	YES	YES	YES
clocks	NO	NO	YES	YES	RESTRICTED	YES	RESTRICTED	YES	YES
random variables	NO	NO	NO	NO	EXP. DIST.	YES	EXP. DIST.	YES	YES
delay nondeterminism	NO	NO	YES	YES	NO	NO	NO	NO	YES
action nondeterminism	YES	YES	YES	YES	NO	NO	YES	YES	YES

TA: Timed automata are transition systems incorporating an explicit notion of real time, represented by continuously moving clocks. Reachability analysis and model checking are the usual techniques employed for TA [1,18]. Timed automata (with deadlines) arise from MoDeST by abstaining from the use of random variables and **palt**.

PTA: Probabilistic timed automata are integrating TA and PTS, thus they arise from STA if random variables are unused. Reachability analysis and model checking have been proposed for PTA [25].

MC: Continuous time Markov chains are a standard model in contemporary performance evaluation. An MC is stochastic process where each delay is governed by some exponential distributed random variable. Analysis techniques for MCs range from the numerical computation of transient and steady state probabilities to approximate model checking [33,2]. MoDeST allows one to model MC by using clocks and exponentially distributed random variables, but in a restricted form (guards are right-continuous and clocks can be uniquely mapped on the random variables they use). Action and delay nondeterminism is not allowed. The model is not closed w.r.t. the operators of the language, e.g. the parallel composition of two MCs is not necessarily a MC (but an IMC).

IMC: Interactive Markov chains are MCs where action nondeterminism can occur. Therefore the model is closed w.r.t. the operators of MoDeST. An IMC can be analysed with algorithms developed for continuous time Markov decision processes [30], or sometimes be reduced to a MC by factoring the model with respect to a weak equivalence [19]. As with MCs these models can be reconstructed from a given STA, if the latter obeys certain restrictions. MoDeST provides shorthand notations making it possible to ensure these restrictions by default: A specification where stochastic aspects only make use of these shorthands possesses a direct semantics in terms of IMC (without reconstructing the latter from the STA semantics).

GSMP: Generalized semi-Markov processes are a general purpose performance evaluation model. Theses stochastic processes are usually analysed using discrete event simulation, but in specific cases a numerical analysis is also feasible. GSMPs arise from MoDeST specifications if action and delay nondeterminism does not occur. The model is not closed w.r.t. the operators of the language, e.g. the parallel composition of two GSMPs is not necessarily an GSMP (but a SA).

SA: Stochastic automata are basically GSMPs with action nondeterminism (hence they are closed under composition), but can also be seen as TA where delay nondeterminism is replaced by random variables governing the delays [13, 11]. As with IMC, specific shorthands can be used to ensure the restrictions required to obtain a SA. For instance if X is a random variable then $\mathsf{wait}(X)$ is an abbreviation for $[x = X \, ; c = 0]$ $\mathsf{urgent}(c \geq x)$ $\mathsf{when}(c \geq x)$ tau where c

(respectively x) is a clock variable (float variable) private to X. Again, these shorthands are used to map the MoDeST specification directly on the SA which otherwise is retrievable from the STA semantics.

It is important to remark that the presence of each listed semantic concept – apart from action nondeterminism – can be detected syntactically, while parsing a specification. This is trivial for probabilistic branching (**palt**), and obvious for clocks, because they have to be declared before use in MoDeST. Use of random variables is easily detected while parsing because (exponential or general) continuous probability distributions are provided via a predefined class (i.e., type). Delay nondeterminism is absent in a specification if for each action the guard and deadline agree. So, Table 2 also gives sufficient syntactic criteria for identifying submodels while parsing a MoDeST specification.

Action nondeterminism is a principal feature for compositional formalisms, yet it induces that MCs and GSMPs are not closed under composition in general. Action nondeterminism can in principle be excluded syntactically by disallowing **alt** and **par**, but the resulting language is too meager to be of much use. More liberal syntactic conditions for absence of action nondeterminism can be adopted from [28].

5   Model Analysis

The identification of well-studied submodels is of crucial practical relevance, because the enormous expressiveness of MoDeST comes with the drawback that the underlying general model is not well investigated: So far analysis methods for the general STA model have not been devised, and their development is ongoing work. The general idea behind this work is strongly based on the identification of submodels of STA for which analysis methods have been published. Based on this knowledge, four different strands can be pursued:

- Isolate syntactic subclasses of MoDeST that map on well-investigated submodels. As long as the user of MoDeST adheres to such a subset, the proper analysis engine can be determined mechanically.
- Define abstractions from STA to less specific models. One such abstraction [12] is to mask the distributions of random clocks i.e., to consider random clocks as delay nondeterministic clocks. In this way, any STA can be turned into a TA by abstracting the stochastic behaviour. Real-time model checking on this TA is safe w.r.t. to the original STA model.
- Define concretisations from more general models to more specific models. This usually means to add additional explicit modelling assumptions, such as to assume a particular scheduler to resolve action nondeterminism, or to assume that all random clocks follow an exponential or phase-type distribution. Note that the quantitative error introduced by such an assumption can be unbounded in certain circumstances.
- Extend or combine analysis methods from submodels of STA to full STA. In particular we are planning to integrate real-time model checking of TA with numerical recipes for GSMPs.

6 Conclusion

In this paper we have introduced a modelling and description language for stochastic timed systems. We have formally defined syntax and semantics of MoDeST, and have put the language in the context of other well-studied models. The focus of this paper has been the behavioural part of MoDeST. The data part is described in [22]. In a nutshell, we allow simple and structured data types, and modularization (packages). Object-oriented enhancements (classes, sub-typing, polymorphism) are under development.

We are currently implementing a tool suite to support modeling and analysis with MoDeST. The language parser is being finalised, and we are working on the state space generator now. The main strategy we pursue in this respect is to bridge to state-of-the-art verification and analysis tools on the level of the STA model. More concretely, we are busy with linking to UPPAAL [27] for real-time model checking and to MÖBIUS [8] for discrete event simulation and numerical analysis.

Acknowledgement The authors are grateful to Ed Brinksma for inspiring discussions. This work is supported by grant TES-4999 of the Dutch Technology Foundation (STW) and grant 612.069.001 of the Netherlands Organisation of Scientific Research (NWO).

A Further MoDeST Expressions

MoDeST provides operations which are shorthands for some common constructions. For instance, both **alt** and **do** allow an **else** alternative (as in Promela). **else** is a shorthand that can be calculated at compile time, e.g.,

$$\text{alt}\{::\text{when}(b_1)\ P_1\ \ldots\ ::\text{when}(b_k)\ P_k\ ::\text{else}\ Q\}$$
$$\stackrel{\text{def}}{=} \text{alt}\{::\text{when}(b_1)\ P_1\ \ldots\ ::\text{when}(b_k)\ P_k\ ::\text{when}(\neg \bigvee_{i=1}^{k} b_i)\ Q\}.$$

In a probabilistic alternative, either assignments or processes (but not both) can be omitted, e.g., $act\ \textbf{palt}\ \{:1:\ [y = 3]\ :2:\ PN(4)\ \}$ should be interpreted as $act\ \textbf{palt}\ \{:1:\ [y = 3]\ \sqrt{}\ :2:\ [\ ]\ PN(4)\ \}$. Notice however that, strictly speaking, the last process is not a legal MoDeST expression since $\sqrt{}$ is not in the language. The following shorthands for assignment are also allowed in MoDeST:

$$[x_1 = e_1,\ \ldots,\ x_k = e_k]\ \stackrel{\text{def}}{=}\ \text{urgent}(\textbf{true})\ \text{tau}\ \textbf{palt}\ \{:\ 1:\ [x_1 = e_1,\ \ldots,\ x_k = e_k]\sqrt{}\}$$
$$x = e\ \stackrel{\text{def}}{=}\ [x = e].$$

Furthermore, invariants like in safety timed automata [18] can be defined by

$$\text{invariant}(b)P\ \stackrel{\text{def}}{=}\ \text{urgent}(\neg b)\text{when}(b)P.$$

MoDeST also provides other useful forms of relabelling apart from **relabel** and **hide**, and standard programming constructs are provided, such as:

$$\text{while}(b)\{P\}\ \stackrel{\text{def}}{=}\ \text{do}\{::\text{when}(b)\ P\ ::\text{else break}\}.$$

References

1. R. Alur and D. Dill. A theory of timed automata. *Th. Comp. Sc.*, **126**:183–235, 1994.
2. C. Baier, J.-P. Katoen, and H. Hermanns. Approximate symbolic model checking of continuous-time Markov chains. In: J.C.M. Baeten and S. Mauw, eds, *Concurrency Theory*, LNCS 1664, pp. 146–161. Springer-Verlag, 1999.
3. G. Berry. Preemption and concurrency. In: R.K. Shyamasundar, ed, *Found. of Software Techn. and Th. Comp. Sc.*, LNCS 761, pp. 72–93. Springer-Verlag, 1993.
4. L. Blair, T. Jones, and G. Blair. Stochastically enhanced timed automata. In: S.F. Smith and C.L. Talcott, eds, *Proc. 4th IFIP Conf. on Formal Methods for Open Object-based Distributed Systems (FMOODS'00)*, pp. 327–347. Kluwer, 2000.
5. T. Bolognesi and E. Brinksma. Introduction to the ISO Specification Language LOTOS. *Computer Netw. and ISDN Sys.*, **14**:25–59, 1987.
6. S. Bornot and J. Sifakis. An algebraic framework for urgency. *Inf. and Comp.*, **163**:172–202, 2001.
7. M. Bravetti and Gorrieri. The theory of interactive generalized semi-Markov processes. *Th. Comp. Sc.*, **258**, 2001 (to appear).
8. D. Daly, D.D. Deavours, J.M. Doyle, P.G. Webster, and W.H. Sanders. Möbius: An extensible tool for performance and dependability modeling. In B.R. Haverkort, H.C. Bohnenkamp, and C.U. Smith, eds, *Computer Performance Evaluation*, LNCS 1786, pp. 332–336. Springer-Verlag, 2000.
9. L. de Alfaro. *Formal Verification of Probabilistic Systems*. PhD thesis, Stanford University, 1997.
10. L. de Alfaro, T.A. Henzinger and R. Majudmar. Stochastic modules. Unpublished manuscript, 1999.
11. P.R. D'Argenio. *Algebras and Automata for Timed and Stochastic Systems*. PhD thesis, Faculty of Computer Science, University of Twente, 1999.
12. P.R. D'Argenio. A compositional translation of stochastic automata into timed automata. Technical Report CTIT 00-08, Faculty of Computer Science, University of Twente, 2000.
13. P.R. D'Argenio, J.-P. Katoen, and E. Brinksma. An algebraic approach to the specification of stochastic systems (extended abstract). In: D. Gries and W.-P. de Roever, eds, *Proc. IFIP Working Conf. on Programming Concepts and Methods*, pp. 126–147. Chapman & Hall, 1998.
14. D. Ferrari. Considerations on the insularity of performance evaluation. *IEEE Trans. on Soft. Eng.*, **12**(6): 678–683, 1986.
15. H. Garavel and M. Sighireanu. On the introduction of exceptions in E-LOTOS. In: R. Gotzhein and J. Bredereke, eds, *Formal Description Techniques IX*, pp. 469–484. Kluwer, 1996.
16. J. Gosling, B. Joy, and G. Steele. *The Java Language Specification*. Addison-Wesley, 1996.
17. C. Harvey. Performance engineering as an integral part of system design. *Br. Telecom Technol. J.*, **4**(3): 142–147, 1986.
18. T.A. Henzinger, X. Nicollin, J. Sifakis, and S. Yovine. Symbolic model checking for real-time systems. *Inf. and Comp.*, **111**:193–244, 1994.
19. H. Hermanns. *Interactive Markov Chains*. PhD thesis, University of Erlangen-Nürnberg, 1998.
20. C.A.R. Hoare. *Communicating Sequential Processes*. Prentice-Hall, 1985.
21. G.J. Holzmann. *Design and Validation of Computer Protocols*. Prentice-Hall, 1991.

22. R. Klaren, P.R. D'Argenio, J.-P. Katoen, and H. Hermanns. Modest language manual. CTIT Tech. Rep. University of Twente, 2001. To appear.
23. P.R. D'Argenio, H. Hermanns, J.-P. Katoen, and R. Klaren. MoDeST – a modelling and description language for stochastic timed systems. CTIT Tech. Rep., University of Twente, 2001.
24. J. Kramer and J. McGee. *Concurrency: State Models and Java Programs*. John Wiley and Sons, 1999.
25. M. Kwiatkowska, G. Norman, R. Segala, and J. Sproston. Automatic verification of real-time systems with probability distributions. In: J.-P. Katoen, ed, *Formal Methods for Real-Time and Probabilistic Systems*, LNCS 1601, pp. 75–95. Springer-Verlag, 1999.
26. M.Z. Kwiatkowska, G. Norman, R. Segala, and J. Sproston. Verifying quantitative properties of continuous probabilistic timed automata. In C. Palamadessi, ed, *Concurrency Theory*, LNCS, Springer-Verlag, 2000.
27. K.G. Larsen, P. Pettersson, and W. Yi. UPPAAL in a nutshell. *Int. J. of Software Tools for Technology Transfer*, 1(1/2):134–152, 1997.
28. V. Mertsiotakis. *Approximate Analysis Methods for Stochastic Process Algebras*. PhD thesis, University of Erlangen-Nürnberg, 1998.
29. J.F. Meyer, A. Movaghar, and W.H. Sanders. Stochastic activity networks: Structure, behavior and application. In: *Proc. Int. Workshop on Timed Petri Nets*, pp. 106–115, IEEE CS Press, 1985.
30. M.L. Puterman. *Markov Decision Processes: Discrete Stochastic Dynamic Programming*. John Wiley & Sons, 1994.
31. R. Segala. *Modeling and Verification of Randomized Distributed Real-Time Systems*. PhD thesis, Dept. of Electrical Eng. and Computer Science, MIT, 1995.
32. A.N. Shiryaev. *Probability*, volume 95 of *Graduate Texts in Mathematics*. Springer-Verlag, 1996.
33. W. Stewart. *Introduction to the Numerical Solution of Markov Chains*. Princeton University Press, 1994.
34. W. Yi. Real-time behaviour of asynchronous agents. In: J.C.M. Baeten and J.-W. Klop, eds, *CONCUR 90*, LNCS 458, pp. 502–520. Springer-Verlag, 1990.

Randomization Helps in LTL Model Checking*

Luboš Brim, Ivana Černá, and Martin Nečesal

Department of Computer Science, Faculty of Informatics
Masaryk University Brno, Czech Republic
{brim,cerna,xnecesal}@fi.muni.cz

Abstract. We present and analyze a new probabilistic method for automata based LTL model checking of non-probabilistic systems with intention to reduce memory requirements. The main idea of our approach is to use randomness to decide which of the needed information (visited states) should be stored during a computation and which could be omitted. We propose two strategies of probabilistic storing of states. The algorithm never errs, i.e. it always delivers correct results. On the other hand the computation time can increase. The method has been embedded into the SPIN model checker and a series of experiments has been performed. The results confirm that randomization can help to increase the applicability of model checkers in practice.

1 Introduction

Model checking is one of the major recent success stories of theoretical computer science. Model checkers are tools which take a description of a system and a property and automatically check whether the system satisfies the property. There are now many different varieties of model checkers including model checkers for real-time systems and probabilistic systems.

Practical application of model checking in the hardware verification became a routine. Many companies in the hardware industry use model checkers to ensure the quality of their products. With the debugging potential afforded by model checking, design of hardware components can be made much more reliable and moreover model checking is seen to accelerate the design process, significantly decreasing the time to market. However, the situation in software model checking is completely different. Software is much more complicated system due to its size and dynamic nature. To achieve similar benefits as in hardware verification, additional methods and techniques need to be explored.

One of the very successful techniques is randomization. The term "probabilistic model checking" (or "probabilistic verification") refers to a wide range of techniques. There are two ways in which probability features in this area. The first approach concerns applying model checking to systems which inherently include probabilistic information [11,4,1,2]. The second approach concerns systems which are non-probabilistic, but of size which makes exhaustive checking

---

* This work has been partially supported by the Grant Agency of Czech Republic grants No. 201/00/1023 and 201/00/0400.

L. de Alfaro and S. Gilmore (Eds.): PAPM-PROBMIV 2001, LNCS 2165, pp. 105–119, 2001.

impractical or infeasible [9,5]. The aim is to use randomization to make model checking more efficient, albeit at a cost of establishing satisfaction with high probability, possibly with a one-sided error, rather than certainty, or at a cost of other resources. While the topic of verification of probabilistic systems has been intensively studied, there are only a few attempts to use randomization in verification of non-probabilistic systems.

In the paper we focus on automata based LTL model checking of non-probabilistic systems. Our aim is to attack the *state-explosion* problem (the number of reachable states grows exponentially in the number of concurrent components and is the main limitation in practical applications of model checkers). Various techniques and heuristics reducing the random access memory required have been proposed. One possible solution (called *on-the-fly* model checking) is to generate only the part of the state graph required to validate or disprove the given property. On-the-fly algorithms generate the state space in a depth-first manner and keep only track of reached states to avoid doing unnecessary work. Another solution makes use of the fact that one of the reasons of the state explosion problem is the generation of all interleavings of independent transitions in different concurrent components. *Partial order reduction* techniques were introduced to ensure that many of these unnecessary interleavings are not explored during state generation.

If we have some knowledge about the structure of the state graph in advance (before starting the actual verification), we can apply even more efficient heuristics. As in general it is not the case we suggest to use a probabilistic method which can be viewed as a probability distribution on a set of deterministic techniques. We explore two probabilistic approaches to achieve significant space reduction in the depth first search based model checking of non-probabilistic systems.

The core of the first approach is to use randomness to decide which of the needed information (visited states) should be stored during a computation and which could be omitted. Consequently, the time complexity of the computation can increase. The second method simply implements the idea of randomizing the branching structure. Both methods are of Las Vegas type, i.e. they always deliver the correct answer. In the paper we focus on the first method and report on the second one briefly. We stress that both methods are compatible (can be used simultaneously) with on-the-fly and partial order reduction techniques. We have implemented both methods and the experiments gave surprisingly very good results in competition with non-probabilistic approaches.

The paper is organized as follows. We first review some background on model checking using automata, define the corresponding graph-theoretic problem, and briefly discuss possible sources for applying randomization. Then we propose the probabilistic reduction algorithm and report experimental results achieved. We conclude with the description of the second method and with some final remarks.

2   Problem Setting

We consider the following verification problem. A finite state transition graph (also called a Kripke structure) is used to represent the behavior of a given system and a linear temporal logic (LTL) formula is used to express the desired property of the system. The basic idea of automata-based LTL model checking is to associate with each LTL formula a Büchi automaton that accepts exactly all the computations that satisfy the formula. If we consider a Kripke structure to be a Büchi automaton as well, then the model checking can be described as a language containment problem and consequently as a non-emptiness problem of (intersecting) Büchi automata. A Büchi automaton accepts some word iff there exists an accepting state reachable from the initial state and from itself. Hence, we can sum up the model checking problem we consider as the following graph-theoretic problem.

Non-emptiness problem of Büchi automata.
Given a directed graph $G = (V, E)$, start state (vertex) $s \in V$, a set of accepting states $F \subseteq V$, determine whether there is a member of F which is reachable from s and belongs to a nontrivial strongly connected component of G.

The direct approach to solve the problem is to decompose the graph into nontrivial strongly connected components (SCCs), which can be done in time linear in the size of the graph using the Tarjan's algorithm [10]. However, constructing SCCs is not memory efficient since the states in the SCCs must be explicitly stored during the procedure. Courcoubetis et al. [3] have proposed an elegant way to avoid the explicit computation of SCCs. The idea is to use a *nested depth-first search* to find accepting states that are reachable from themselves (to compute *accepting path*). The pseudo-code of the NestedDFS algorithm is given in Fig. 1. Only two bits need to be added to each state to separate the states stored in *VisitedStates* during the first and the second (nested) DFS. The extreme space efficiency of the NestedDFS algorithm is achieved to the detriment of time. The time might double when all the states are reachable in both searches and there are no accepting cycles. However, in applications to real systems the space is actually more critical resource. This makes the nested depth-first search the main algorithm used in many verification tools which support the automata based approach to model checking of LTL formulas (e.g. SPIN).

The space requirements of the NestedDFS algorithm are determined by the necessity of storing *VisitedStates* in randomly accessed memory. Several implementations of NestedDFS use different data structures to represent the set *VisitedStates*. The basic one is a *hash table* [6]. Another implementation [12] makes use of symbolic representation of *VisitedStates* via *Ordered Binary Decision Diagrams* (OBDD).

Hash compaction is used in [14], where the possible hash-collisions are not re-solved. The algorithm can thus detect a state as visited even if it is not. Consequently, not all reachable states are explored during the search, and an error might go undetected.

```
proc DFS(s)
    add {s,0} to VisitedStates;
    foreach successor t of s do
        if {t,0} not in VisitedStates then DFS(t) fi
    od;
    if accepting(s) then seed := s; NDFS(s) fi
end
```

```
proc NDFS(s)
    add {s,1} to VisitedStates;
    foreach successor t of s do
        if {t,1} not in VisitedStates
            then NDFS(t)
            else if t = seed then "report cycle" fi fi
    od
end
```

Fig. 1. Algorithm NestedDFS

Another technique which has been investigated to reduce the amount of randomly accessed memory is *state-space caching* [5]. The idea is based on the observation that when doing a depth-first search of a graph, storing only the states that are on the search stack is sufficient to guarantee that the search terminates. While this can produce a very substantial saving in the use of randomly accessed memory, it usually has a disastrous impact on the run time of the search. Indeed, each state will be visited as many times as there are simple paths reaching it. An improvement on this idea is to store not only the states that are on the search stack, but also a bounded number of other states (as many as will fit into the chosen "state-space cache"). If the state-space cache is full when a new state needs to be stored, random replacement of a state that is not currently on the search stack is used.

The advantage of state-space caching is that the amount of memory that is used can be reduced with a limited impact on the time required for the search. Indeed, if the cache is large enough to contain the whole state space, there is no change in the required time. If the size of the cache is reduced below this limit, the time required for the search will only increase gradually. Experimental results, however, show that below a threshold that is usually between 1/2 and 1/3 of the size of the state space, the run time explodes, unless additional techniques are used to restrict the number of distinct paths that can reach a given state [5].

The behavior of state-space caching is quite the opposite of that of the hashing technique. Indeed, state-space caching guarantees a correct result, but at the cost of a potentially large increase in the time needed for the state-space search. On the other hand, hashing never increases the required run time, but can fail to

```
proc DFS(s)
    add {s,0} to VisitedStates;
    foreach successor t of s do
        if {t,0} not in VisitedStates then DFS(t) fi
    od;
    if accepting(s) then seed := s; NDFS(s) fi;
    if ReductionStrategy(s) then delete {s,*} from VisitedStates fi
end


proc NDFS(s)
    if accepting(s) and {s,0} not in VisitedStates then exit fi;        (×)
    add {s,1} to VisitedStates;
    foreach successor t of s do
        if {t,1} not in VisitedStates
            then NDFS(t)
            else if t = seed then "report cycle" fi fi
    od;
    if {s,0} not in VisitedStates then delete {s,1} from VisitedStates fi
end
```

Fig. 2. Algorithm NestedDFSReSt

explore the whole state space. A combination of state space caching and hashing has been proposed and investigated in [9].

In this paper we propose a new technique to attack the state-explosion problem using a simple *probabilistic method*. Actually, the technique has been strongly motivated by our intention to improve the performance of the model checker SPIN, and the technique has been embedded into SPIN for testing purposes.

The proposed method allows to solve the emptiness problem of Büchi automata (i.e. complete LTL model checking and not only reachability) and it never errs. It can be briefly described in the following way. The algorithm is based on the nested depth-first search as described in Fig. 1. Each time the algorithm backtracks through a state it employs a proper reduction strategy to decide whether the state will be kept in the *VisitedStates* table or whether it will be removed. We propose two reduction strategies, the dynamic and the static one. While the first one takes on the frequency of visiting the state, the second one allows to eliminate delayed storing of the state and thus decreases the number of visits of individual states. We specify properties of systems determining which strategy suits better for a given verification problem.

3   Algorithm with Probabilistic Reduction Strategy

The reason to store the states in the table of visited states during nested depth-first search is to speed up the verification by preventing the multiplication of work when states are re-visited. A state that is visited only once need not be stored at all, while storing a state which will be visited many times can result in a significant speed-up. The standard nested depth-first search algorithm stores *all* visited states. On the other side, the optimal strategy for storing states would take into account the number of times a state will be eventually visited – a *visitation factor*. As it is generally impossible to compute this parameter in advance, we will use probabilistic method to solve the problem.

The pseudo-code of the modified nested depth-first-search algorithm with reduction strategy, NestedDFSReSt, is given in Fig. 2. Whenever the *DFS* procedure explores a new state, the state is temporally saved in the *VisitedStates* table (with parameter 0). Whenever *DFS* backtracks trough a state, a test *ReductionStrategy* is performed and if the test evaluates to *true* the state is *removed* from the *VisitedStates* table. We will consider two basic probabilistic strategies of removing states. The first one *dynamically* decides on removing a state each time the state is backtracked through, while the second heuristic decides randomly in advance (before the verification is started) which states will be stored permanently.

As in the case of *DFS*, the *NDFS* procedure also needs the list of states it has visited to be efficient. Therefore every exploring of a new state results in its saving to the *VisitedStates* table (with parameter 1). Whenever *NDFS* backtracks trough a state it respects the *ReductionStrategy* test performed on this state by the *DFS* procedure and if necessary removes the state from the table.

Removing states from the *VisitedStates* table has direct impact on the time complexity of the algorithm as re-visiting a state removed from the table invokes a new search from this state.

The correctness of the NestedDFSReSt algorithm follows from the correctness of the NestedDFS algorithm [3]. The additional key arguments it depends on are summarized in the following two lemmas.

Lemma 1. *During the whole computation the sequence of states with which the DFS procedure is called (DFSstack) forms a path in the graph G. The same is true for the NDFS procedure and NDFSstack.*

Proof: The *(N)DFS* procedure is always called with the argument t which is a successor of the current state s.

Lemma 2. *Suppose that during the whole computation both the DFSstack and the NDFSstack are subsets of* VisitedStates, *then the* NestedDFSReSt *algorithm terminates.*

Proof: From the inclusion follows that the *(N)DFSstack* always forms a simple path. The number of simple paths in G is finite and each one is explored at most once.

Theorem 1. *The algorithm* NestedDFSReSt *is correct.*

Proof: Whenever the *(N)DFS* procedure explores a new state, the state is temporally saved in the *VisitedStates* table. Therefore (N)DFSstack $\subseteq$ *VisitedStates* is invariantly true and NestedDFSReSt always terminates due to the Lemma 2. If NestedDFSReSt reports "cycle" then due to the Lemma 1 there is a reachable cycle containing an accepting state. Conversely, suppose there is a reachable cycle containing an accepting state in G. Deleting states from *VisitedStates* table cannot cause leaving out any call of *(N)DFS(t)* which would have been performed by NestedDFS algorithm. Moreover, the situation in which the condition of the **if** test on the very first line (denoted by ×) in *NDFS* is true is equivalent to the situation when $\{s, 1\}$ is in *VisitedStates* in NestedDFS algorithm. Therefore NestedDFSReSt searches trough all the paths NestedDFS does and thus reports "cycle" when NestedDFS does. ∎

Notice that the test on the first line (×) of *NDFS* prevents re-searching of an accepting state and thus speeds-up significantly the overall time complexity. This fact was confirmed also by experimental results.

The proof of the Theorem 1 is based on the fact that the NestedDFSReSt algorithm searches through all the paths the NestedDFS one does. Due to this fact our algorithm is compatible with additional techniques used for state space reductions, especially with partial order reduction techniques used in SPIN.

3.1 Dynamic Reduction Strategy

The pseudo-code implementing the dynamic reduction strategy is as follows:

funct *ReductionStrategy-Dynamic(s)* : *boolean*
  $p := random[0, 1]$;
  if $p \leq P_{del}$
    then *ReductionStrategy-Dynamic* := *true*
      else *ReductionStrategy-Dynamic* := *false* **fi**
end

P_{del} is a fixed parameter determining the probability of deleting a state from *VisitedStates* table. Each time the *DFS* backtracks through a state s the state is deleted with the probability P_{del} and is kept stored with the probability $P_{sto} = 1 - P_{del}$. Once a state is kept stored in the table by the *DFS* procedure, it is never removed. The probability that a state will be eventually stored thus depends on the number k of its visits during the computation and is equal to $Prob(s$ is eventually stored$) = 1 - P_{del}^k$. This means that a state with higher visitation factor k has also higher probability to be stored permanently. The probability that the state s will be re-visited more than i times is equal to $Prob(s$ is i times deleted$) = P_{del}^i$.

The dynamic reduction strategy would lead to a non-trivial reduction of randomly accessed memory if there is a non-trivial subset of the state space that will never be permanently stored. The expected memory reduction can be expressed as $P \times (size\ of\ the\ state\ space)$, where P is the probability that a state will never be permanently stored. If k is the average visitation factor then P can be estimated as P_{del}^k. Therefore, we would like to have the highest possible value of the probability that a state will never be permanently stored.

On the other hand, not saving a frequently visited state increases the time complexity of the whole computation. Therefore, we are interested in the expected number of visits after which the state is stored permanently. Consider an elementary event $\{s$ is permanently stored during its i-th visit$\}$. Then

$$Prob(\{s \text{ is permanently stored during its } i\text{-th visit}\}) = P_{del}^{i-1}P_{sto}.$$

Let H be a random variable over the above mentioned elementary events defined as

$$H(\{s \text{ is permanently stored during its } i\text{-th visit}\}) = i.$$

We have that the expected value of H is

$$E(H) = \sum_{i=1}^{\infty} i P_{del}^{i-1} P_{sto} = P_{sto} \sum_{i=1}^{\infty} i P_{del}^{i-1} = P_{sto} \sum_{j=1}^{\infty}\sum_{i=j}^{\infty} P_{del}^{i-1} =$$

$$= P_{sto} \sum_{j=1}^{\infty} \frac{P_{del}^{j-1}}{1 - P_{del}} = \frac{P_{sto}}{1 - P_{del}} \sum_{j=0}^{\infty} P_{del}^{j} = \frac{P_{sto}}{1 - P_{del}} \frac{P_{del}^{0}}{1 - P_{del}} =$$

$$= \frac{P_{sto}}{(1 - P_{del})^2} = \frac{P_{sto}}{P_{sto}^2} = \frac{1}{P_{sto}}$$

It can be seen that the expected value of the random variable H depends on the probability P_{sto} and indicates that value P_{sto} should be high.

We can conclude that in systems with a high visitation factor we cannot expect reasonable space savings without enormous increase of the time complexity.

3.2 Static Reduction Strategy

The second strategy tries to eliminate the main disadvantage of the dynamic reduction strategy, namely the delayed storage of a state. If a state will eventually be permanently stored, why not to store it immediately during the first visit. When deciding which states are to be stored we should prefer states with high visitation factor. As we cannot compute this factor in advance we use probabilistic decision. All states are *in advance* and *randomly* divided into two groups: states which will be stored and those which will never be stored (represented as R). Hence, each state is permanently stored during its first visit or never. The ratio between stored and non-stored states is selected with the intention to achieve as highest reduction in state space as possible.

The pseudo-code implementing the static reduction strategy is as follows:

```
funct ReductionStrategy-Static(s) : boolean
  if s ∈ R
    then ReductionStrategy-Static := true
    else ReductionStrategy-Static := false fi
end
```

The disadvantage of the static reduction strategy is its insensibility to the visitation factor.

4 Experiments

To be able to compare experimentally our probabilistic algorithm with the non-probabilistic one, we have embedded the algorithm into SPIN model checker.

We have performed a series of tests on several types of standard parametrized (scalable) verification problems. Here we report on two of them only:

Peterson. Peterson's algorithm solves the mutual exclusion problem. We have considered the algorithm for parameter $N = 3$ determining the number of processes. The property to be verified was $\Box(ncrit < 2)$ (no more than one process is in critical section).

Philosophers. Dining Philosophers is a model of a problem of sharing of resources by several processes. We have considered the algorithm for $N = 4$ and $N = 6$. The property to be verified was $\Box\Diamond(EatingAny = 1)$ (absence of deadlock).

The other problems we have considered were e.g. the Leader Election problem, Mobile processes. In all these experiments we have obtained similar results.

As our algorithm is compatible with partial order reduction techniques used in SPIN we have compiled all problems with partial order reductions.

For each verification problem we first give two most important characteristics of the computation performed by SPIN checker: *States* (the number of states saved in the *VisitedStates* table) and *Transitions* (the number of performed transitions). The number of transition is proportional to the overall time complexity of the computation. The size of the *VisitedStates* table in SPIN's computation is nondecreasing. Once a state is stored in the table it is never removed. On the other hand in the NestedDFSReSt algorithm every visited state is temporally stored in the table and only when it is backtracked through the (random) decision about its permanent storing is made. Therefore for our algorithm we need another characteristic, namely the highest size of the *Visited States* table, *Peak States*. The parameter *States* declares the number of states stored in the table at the end of computation. The remaining two parameters, *State Saving* and *Transition Overhead*, compare performance of the deterministic algorithm and the probabilistic ones. Computations of probabilistic algorithms were repeated 10 times, presented values are the average ones.

Peterson's Algorithm

Results of experiments are summarized in the Table 1. The best results with Dynamic Strategy were achieved for storing probability 0.5 where saving in the size of stored state space was 33% while increase in the time was negligible, and for probability 0.1 with 52% space saving and multiplication factor 4 of time. To get deeper inside we mention that the computation without the reduction strategy took about 1.5 second in this case. Yet another increase in the deleting probability results in substantial grow of time but does not improve space saving factor significantly.

Table 1. Summary of Experimental Results for **Peterson**

	States	Peak	Saving	Transitions	Overhead
SPIN	17068	17068	0%	32077	1.00
Dynamic Strategy					
$P_{sto} = 0.50$	10998	11421	33%	46074	1.44
$P_{sto} = 0.10$	6724	8263	52%	136344	4.25
$P_{sto} = 0.01$	5559	7407	57%	1110526	34.62
Static Strategy					
$P_{sto} = 0.75$	12807	12812	25%	38761	1.21
$P_{sto} = 0.50$	8568	9661	43%	63662	1.98
$P_{sto} = 0.40$	6852	8417	51%	390737	12.18

Experiments with Static Strategy reveal that we can achieve 43% space saving for the price of double time complexity. 51% space saving is attained with worse time multiplication factor (12 in comparison to 4) than in the case of Dynamic Strategy. The difference between storing probability and real space savings (i.e. for storing probability 0.4 we would expect 60% saving instead of measured 51%) has two reasons. Firstly, as we do not know which states of the state space are actually reachable in the verified system we have to divide the whole state space in advance. Secondly, the division determines states which are permanently saved but the *VisitedStates* table contains also temporally saved states and its size can be temporally greater (parameter *Peak States*). State space saving is computed via comparing the number of saved states by non-probabilistic computation and the peak value of probabilistic computation.

Dining Philosophers

Results of experiments are summarized in the Table 2 for $N = 4$ and in the Table 3 for $N = 6$. In both cases the results are comparable. Dynamic Strategy

Table 2. Summary of Experimental Results for **Philosophers** with $N = 4$

	States	Peak	Saving	Transitions	Overhead
SPIN	3727	3727	0%	18286	1.00
Dynamic Strategy					
$P_{sto} = 0.50$	3047	3178	15%	33475	1.83
$P_{sto} = 0.10$	2482	2686	28%	139263	7.62
$P_{sto} = 0.01$	2316	2531	32%	1287156	70.39
Static Strategy					
$P_{sto} = 0.75$	2788	2961	21%	49112	2.69
$P_{sto} = 0.60$	2221	2577	31%	232973	12.74
$P_{sto} = 0.50$	1875	2340	37%	3285607	179.68

again gives the best results for storing probability between 0.5 and 0.1. Any further decrease in the storing probability below 0.1 results in significant increase of time complexity. In the case of Static Strategy reasonable results were obtained for storing probability 0.75 and further decreasing of probability leads to unreasonable time overhead and thus prevents from higher space savings.

Table 3. Summary of Experimental Results for **Philosophers** with $N = 6$

	States	Peak	Saving	Transitions	Overhead
SPIN	191641	191641	0%	1144950	1.00
Dynamic Strategy					
$P_{sto} = 0.50$	160426	165461	14%	2152384	1.88
$P_{sto} = 0.10$	136081	145214	24%	9400300	8.21
$P_{sto} = 0.01$	131306	140758	27%	91533400	79.90
Static Strategy					
$P_{sto} = 0.75$	143661	155920	19%	6702840	5.85
$P_{sto} = 0.65$	124377	143691	25%	116103466	101.40

Generally, the results for Philosophers are worse than those for Peterson's algorithm and are remarkably influenced by the visitation factor. While in the Peterson's algorithm the average number of state visits in SPIN's computation is $32077/17068 = 1.8$, in Philosophers it is 4.9 ($N = 4$) and 6 ($N = 6$). Experimental observations are thus in accordance with deduced theoretical results.

5   Random Nested DFS

Besides the algorithm with probabilistic reduction strategy we have also explored the potential of randomizing the branching points in nested depth first search. Verification tools typically build the state space from the syntactical description of the problem. E.g. in SPIN the **foreach** *successor t of s* **do** in the depth first search is implemented as **for** $i = 1$ **to** n cycle. This means that the search order is fixed by the input PROMELA program describing the system. If the verification fails due to space limitations it is recommended to re-write the program to re-order the guarded commands in conditionals and loops. However, the user typically has no information on what would be a good re-arrangement. Hence, the situation is very suitable for a randomized approach.

We have implemented the **foreach** *successor t of s* **do** in the depth first search as a *random selection* of the successors ordering and performed a series of comparisons with the standard SPIN tool on similar set of problems as we did before. Even though the method is trivial, the results we obtained were quite surprising. For instance for the **Philosophers** (with an error) the results are partially summarized in the Table 4.

Table 4. Summary of Experimental Results for **Random Nested DFS**

N	SPIN			Random NDFS				
	States	Trans	Memory	Runs	Success	States	Trans	Memory
11	288922	1449200	56.9 MB	10	10	100421	505150	26.4
12			205.0 MB	10	3	68355	346824	19.9
14			2.8 GB	50	5	46128	250266	16.2
16			38.5 GB	50	5	46288	245406	17.8
20			6.7 TB	50	2	38282	213639	18.2

For the value of the parameter N greater than 11 the SPIN model checker was not able to complete the computation. We therefore give *estimated* values for the memory requirements obtained by extrapolation from finished computations. The randomized algorithm was repeatedly performed (*Runs*) and the number of successful runs (discovering the error before memory overflow) is reported (*Success*). The experiments indicate that even a small number of repetitions can dramatically increase the power of the tool.

We have also considered some artificial verification examples, which demonstrate the potential of the method in some extreme cases. Consider the following verification problem defined by the program

```
1 proc ExIF
2    MainCounter := 0;  StepCounter := 0;
3    while StepCounter < 1000 do
4          if
5             true → MainCounter := MainCounter + 1
6             true → MainCounter := MainCounter + 2
7          fi;
8          StepCounter := StepCounter + 1 od
9 end
```

and the LTL formula

$$\Box(MainCounter < 2000 - Diff)$$

The parameter *Diff* determines the ratio of runs of the program that satisfy and violate the formula. More precisely, the probability that $MainCounter = 2000 - Diff$ is

$$Prob(MainCounter = 2000 - Diff) = \binom{1000}{Diff} \left(\frac{1}{2}\right)^{1000}$$

We have performed experiments for various values of *Diff*. The results are summarized in the Table 5. The experiments have confirmed that the actual memory savings strictly depend on the value of the parameter *Diff*, that is on the probability of a faulty run, and have ranged from 20% up to 90%. We stress that after re-ordering of guarded commands in the *ExIF* program (swapping lines 5 and 6)

SPIN finds the counterexample immediately. Re-writing the program helps in this case. The next example shows that in some situations even re-writing the program does not help.

Table 5. Summary of Experimental Results for **ExIF**

Diff	ViolProb	Algorithm	States	Transitions	%
400	$1.3642.10^{-10}$	RandNestedDFS	37999	55810	10.4%
		SPIN	363202	543502	100.0%
200	$8.2249.10^{-96}$	RandNestedDFS	368766	551537	57.3%
		SPIN	643602	964002	100.0%
100	$6.7017.10^{-162}$	RandNestedDFS	647855	969977	79.6%
		SPIN	813802	1219250	100.0%

Let us consider the LTL formula

$$\Box((StepCounter < 1000) \lor (MainCounter \neq 1500)).$$

The formula expresses the property that at the end of every computation (i.e. when $StepCounter = 1000$) the value of $MainCounter$ is not 1500. It is easy to see that the $ExIF$ program does not fulfil this property. The erroneous computations are those where both guards are selected equally. For every re-ordering of the guards SPIN has to search the significant part of the state space to discover a counterexample. On the other hand, the $RandNestedDFS$ algorithm successes very quickly as it selects both guards with the same probability. The same effect has been observed in other tests as well. E.g. in the Leader election problem, for every permutation of all guards SPIN has searched approximately the same number of states while $RandNestedDFS$ has needed to search through significantly smaller part of the state space.

6 Conclusions

While verification of probabilistic systems seems to be ready to move to the industrial practice, the use of probabilistic methods in model checking of non-probabilistic systems is at its beginning. The use of probabilistic methods in the explicit state enumeration techniques to reduce the memory required by hashing is an excellent example of the potential of probabilistic methods. Our intention was to investigate other possibilities how randomization could help in model checking.

We have proposed a new probabilistic verification method which could reduce the amount of random access memory necessary to store the information about the system. The reduction rate depends on the verified system, namely on the average number of state visits. Our experiments have confirmed that the method could achieve a non-trivial reduction within reasonable time overhead.

Another important issue for further study is to examine possibilities of combining our probabilistic reduction strategy algorithm with other techniques to reduce memory usage. We also plan to perform additional experiments to give a more comprehensive view of the performance of our technique and of its scalability.

Acknowledgement. We would like to thank Jiří Barnat for introducing us to the mysteries of the SPIN model checker and for his advice and efficient help with incorporating our algorithms into SPIN.

References

1. C. Baier and M. Kwiatkowska. Model Checking for a Probabilistic Branching Time Logic with Fairness. *DISTCOMP: Distributed Computing*, 11, 1998.
2. A. Bianco and L. De Alfaro. Model Checking of Probabilistic and Nondeterministic Systems. In P. S. Thiagarajan, editor, *Foundations of Software Technology and Theoretical Computer Science (FSTTCS)*, volume 1026 of *LNCS*, pages 499–513. Springer-Verlag, 1995.
3. C. Courcoubetis, M. Vardi, P. Wolper, and M. Yannakakis. Memory-Efficient Algorithms for the Verification of Temporal Properties. *Formal Methods in System Design*, 1:275–288, 1992.
4. C. Courcoubetis and M. Yannakakis. The Complexity of Probabilistic Verification. *Journal of the ACM*, 42(4):857–907, July 1995.
5. P. Godefroid, G. J. Holzmann, and D. Pirottin. State-Space Caching Revisited. *Formal Methods in System Design: An International Journal*, 7(3):227–241, November 1995.
6. G. J. Holzmann. An Analysis of Bitstate Hashing. *Formal Methods in System Design: An International Journal*, 13(3):289–307, Nov. 1998.
7. G.J. Holzmann, D. Peled, and M. Yannakakis. On Nested Depth First Search. In *The SPIN Verification System*, pages 23–32. American Mathematical Society, 1996. Proc. of the Second SPIN Workshop.
8. M. Narasimha, R. Cleaveland, and P. Iyer. Probabilistic Temporal Logics via the Modal Mu-Calculus. In W. Thomas, editor, *Proceedings of the Second International Conference on Foundations of Software Science and Computation Structures (FoSSaCS)*, volume 1578 of *LNCS*, pages 288–305. Springer-Verlag, 1999.
9. U. Stern and D.L. Dill. Combining State Space Caching and Hash Compaction. In B. Straube and J. Schoenherr, editors, *4. GI/ITG/GME Workshop zur Methoden des Entwurfs und der Verifikation Digitaler Systeme*, pages 81–90. Shaker Verlag, Aachen, 1996.
10. R. Tarjan. Depth First Search and Linear Graph Algorithms. *SIAM journal on computing*, pages 146–160, Januar 1972.
11. M. Vardi. Probabilistic Linear-Time Model Checking: An Overview of the Automata-Theoretic Approach. In J.-P. Katoen, editor, *International AMAST Workshop on Formal Methods for Real-Time and Probabilistic Systems (ARTS)*, volume 1601 of *LNCS*, pages 265–276. Springer-Verlag, 1999.
12. W. Visser. Using OBDD Encodings for Space Efficient State Storage during On-the-fly Model Checking. Proceedings of the 1st SPIN Workshop, Montreal, Canada, 1995.

13. A. K. Wisspeintner, F. Huber, and J. Philipps. Model Checking and Random Competition – A Study Using the Model Checking Framework MIC. 10. GI/ITG Fachgespräch "Formale Beschreibungstechniken für verteilte Systeme", pages 91–100, June 2000.
14. P. Wolper and D. Leroy. Reliable Hashing Without Collision Detection. In C. Courcoubetis, editor, *Proc. 5th International Computer Aided Verification Conference (CAV'93)*, volume 697 of *LNCS*, pages 59–70. Springer-Verlag, 1993.

An Efficient Kronecker Representation for PEPA Models*

Jane Hillston[1] and Leïla Kloul[2]

[1] LFCS, University of Edinburgh, Kings Buildings, Edinburgh EH9 3JZ, Scotland
jeh@dcs.ed.ac.uk
[2] PRiSM, Université de Versailles, 45, Av. des Etats-Unis, 78000 Versailles, France
kle@prism.uvsq.fr

Abstract. In this paper we present a representation of the Markov process underlying a PEPA model in terms of a Kronecker product of terms. Whilst this representation is similar to previous representations of Stochastic Automata Networks and Stochastic Petri Nets, it has novel features, arising from the definition of the PEPA models. In particular, capturing the correct timing behaviour of cooperating PEPA activities relies on functional dependencies.

1 Introduction

Performance investigation of modern computer and communication systems requires the development of relevant and efficient modelling techniques. The rich synchronisation constraints and the size of these systems lead to complex models with exponential growth of the number of states. Traditional performance models, based on queueing networks, cannot readily capture these constraints; thus several new performance modelling techniques have been developed, e.g. Stochastic Petri Nets (SPN), Stochastic Automata Networks (SAN) and Stochastic Process Algebras (SPA).

Petri nets were designed to represent synchronisation constraints within concurrent systems and protocols; SPN associate random variables with timed transitions within the net [16,1,20]. However, although the graphical representation of Petri Nets presents the dynamic behaviour of the model, it provides little insight into the structure of the system being modelled.

SAN and SPA provide mechanisms which allow the increasing complexity of synchronisation constraints to be captured whilst retaining the compositional structure of the system explicitly within the model. For many modern systems, being able to construct a model from components or elements, reflecting the system's composition, greatly aids handling the complexity of the model construction task. As with all state-based modelling formalisms, such models are prone to state space explosion. However, both formalisms incorporate techniques for overcoming this problem.

* This work is supported by CNRS/RS project (UIIVV 78171) and EPSRC project COMPA (G/L10215).

L. de Alfaro and S. Gilmore (Eds.): PAPM-PROBMIV 2001, LNCS 2165, pp. 120–135, 2001.

The SAN formalism, developed by Plateau [18], models complex systems with interacting components such as parallel systems. To tackle the state space explosion problem, Plateau [17] has proved that the generator matrix of the Markov process underlying a SAN model can be analytically represented using Kronecker algebra. Moreover the solution of the model can be achieved via this tensor expression of submatrices—the complete matrix does not need to be generated.

SPA are extensions of classical process algebras such as CCS and CSP, analogous to SPN in the sense that random variables are associated with timed actions in the model. In this paper we consider the Markovian process algebra, Performance Evaluation Process Algebra (PEPA), introduced by Hillston in 1994 [14]. In PEPA, a system is described as an interaction of components which, either singly or multiply, engage in activities. The components represent the active parts within the system and the activities the actions of those parts.

Various techniques for solving large models have been developed for PEPA but these have focused on aggregation or decomposition techniques, which use the process algebra structure of the model to guide manipulations of the underlying Markov process. In this paper we show that a PEPA model can also be represented analytically using Kronecker algebra and solved without constructing the complete generator matrix. Correct representation of the features of the PEPA model, in particular the synchronisation behaviour, relies on the functional dependencies introduced in PEPA formalism in [15]. Just as for SAN models, we show that the translation from the model to the compact representation is automatic.

This paper is structured as follows. In Sect. 2, we present the PEPA language. A small example illustrates the use of this modelling technique. Section 3 is dedicated to the functional depencies in PEPA. In Sect. 4, we show how to represent the underlying Markov process of a PEPA model using the tensor algebra. An application example is given, followed by the proof of the validity of this analytical representation. Section 5 is dedicated to related work. Finally, we conclude with some remarks and future work.

2 PEPA

The basic elements of PEPA[14] are *components* and *activities*, corresponding to *states* and *transitions* in the underlying Markov process. Each activity has an *action type* and τ denotes the distinguished type representing private or unseen action. The duration of each activity is represented by the parameter of the associated exponential distribution: the *activity rate* of the activity. The rate may be any positive real number, or the distinguished symbol $\top$ (read as *unspecified*). Thus each activity, a, is a pair (α, r) where α is the type and r is the rate. We let $\mathcal{C}$ denote the countable set of components and $\mathcal{A}$ denote the countable set of all possible action types. We denote by $Act \subseteq \mathcal{A} \times \mathbb{R}^+$, the set of activities, where $\mathbb{R}^+$ is the set of positive real numbers plus the symbol $\top$. Models in PEPA are built using a small but expressive set of combinators:

Prefix $(\alpha, r).C_1$. Prefix is the basic mechanism by which the behaviours of components are constructed. The component carries out activity (α, r) and subsequently behaves as component C_1.

Choice $C_1 + C_2$. The component represents a system which may behave either as component C_1 or as C_2: all the current activities of both components are enabled. A *race condition* determines the first activity to complete, and so distinguishes one component; the other is discarded.

Cooperation $C_1 \bowtie_L C_2$. The components proceed independently with any activities whose types do not occur in the *cooperation set* L. However, activities with action types in the set L require the simultaneous involvement of both components; these *shared activities* are only enabled when they are enabled in both C_1 and C_2. The shared activity occurs at the rate of the slowest participant. The capacity of a component C_i to carry out a given action type α (the sum of rates associated with its α actions) is called its *apparent rate*, denoted $r_\alpha(C_i)$. The apparent rate of a shared activity is the minimum, among the participating components, of the apparent rates for that type.

If an activity has rate $\top$ the component is *passive* with respect to that action type and it does not influence the rate at which such shared activities occur. When the set L is empty, we use the more concise notation $C_1 \parallel C_2$ to represent $C_1 \bowtie_\emptyset C_2$.

Hiding C_1/L. The component behaves as C_1 except that any activities of types within the set L are *hidden*, i.e. such an activity exhibits the unknown type τ and the activity can be regarded as an internal delay by the component. The original action type of a hidden activity is no longer accessible; the duration is unaffected.

Constant $M \stackrel{def}{=} C_1$. Constants are components whose meaning is given by a defining equation: $M \stackrel{def}{=} C_1$ gives the constant M the behaviour of the component C_1.

The semantics of PEPA, presented in the structured operational semantics style, are given in [14]. The underlying transition system also characterises the Markov process represented by the model. Rules are given for each of the combinators, showing how the component may evolve. Here we show only the rule for shared activities (Fig. 1).

$$\frac{P \xrightarrow{(\alpha, r_1)} P' \quad Q \xrightarrow{(\alpha, r_2)} Q'}{P \bowtie_L Q \xrightarrow{(\alpha, r)} P' \bowtie_L Q'} \ (\alpha \in L), \qquad r = \frac{r_1}{r_\alpha(P)} \frac{r_2}{r_\alpha(Q)} \min(r_\alpha(P), r_\alpha(Q))$$

Fig. 1. Operational rule defining shared activities

From a model definition M we can apply the semantic rules exhaustively to find the complete set of reachable states, the *derivative set* of M, $ds(M)$. From this set, we can construct the *derivation graph*. The derivation graph is a directed multigraph whose set of nodes is $ds(M)$ and whose arcs represent the possible transitions between them. To derive a Markov process from a PEPA model we associate a state with each node of the derivation graph. Action type information is discarded so that edges are labelled only by rates; multiple edges between a pair of nodes are combined by summing the corresponding rates. The rate on an edge in this modified graph becomes the corresponding entry in the infinitesimal generator matrix. Thus the rate between components C and C' is denoted $q(C, C')$. Similarly the conditional transition rate between C and C' due to activities of type α is denoted $q(C, C', \alpha)$.

Necessary (but not sufficient) conditions for the ergodicity of the Markov process in terms of the structure of the PEPA model have been identified and can be readily checked [14]. These imply that the model should be constructed as a cooperation of *sequential* components, i.e. components constructed using only prefix, choice and constants. Thus the compositional structure of PEPA models is at the level of the cooperating components; we refer to such models as *well-defined*. Syntactic analysis can be used to determine all the action types which will occur within the lifetime of a component C, a set denoted $\mathcal{A}(C)$.

Example. Consider a simple two-place buffer and a server. The buffer accepts arrivals with a rate λ and passes the contents for service. When there are two customers in the buffer each attempts service, but only the front customer can be successfully serviced with rate s. Service of the second customer results in a partial service which must be corrected, at rate t before the buffer can make any other action. The server simply accepts customers for service (passively) and then allows them to depart, carrying out a false departure after a partial service. Let d be the departure rate.

In PEPA, the system is represented as the interaction of two components $Buffer_0$ and $Server$. We use three action types: *in*, *service* and *depart*. The first describes the arrival of a new customer in the buffer, the second, the service completion, and the last one the departure of a customer from the server. The components are defined as shown below.

$$Buffer_0 \stackrel{def}{=} (in, \lambda).Buffer_1 \qquad\qquad Server \stackrel{def}{=} (service, \top).Server'$$
$$Buffer_1 \stackrel{def}{=} (service, s).Buffer_0 + (in, \lambda).Buffer_2 \qquad Server' \stackrel{def}{=} (depart, d).Server$$
$$Buffer_2 \stackrel{def}{=} (service, s).Buffer_1 + (service, s).Buffer_3$$
$$Buffer_3 \stackrel{def}{=} (service, t).Buffer_1$$

In addition to the mutually recursive sets of equations defining the behaviour of each sequential component, we have a *system equation* which defines the cooperation between the two components.

$$System \stackrel{def}{=} Buffer_0 \underset{\{service\}}{\bowtie} Server$$

3   Functional Dependencies

In SAN, automata are able to influence one another in two ways, both related to events. Direct interaction between automata is modelled by synchronised transitions, equivalent to cooperation or shared activities in PEPA.

The other form of interaction is less direct: transition rates within an automaton can be influenced by the local states of one or more other automata of the network. Using such rates may lead to a reduction in the model size since functional rates are a means to avoid explicitly modelling all parts of a system's bahaviour. This benefit is most appreciated when building/solving the underlying Markov chain. In [15], we have introduced the notion of functional dependencies between PEPA model's components by extending the activity rates to include functional rates.

In PEPA the set of activities $\mathcal{Act}$ is defined as $\mathcal{Act} \subseteq \mathcal{A} \times \mathbb{R}^+$ where $\mathbb{R}^+$ is the set of positive real numbers defined as follows:

$$\mathbb{R}^+ = \{r | r > 0; r \in \mathbb{R}\} \cup \{\top\}$$

In the context of PEPA, a functional dependency may involve one or several components. In a functional dependency involving a single component, the rate value of one or several activities of the component depends on the current state of the component itself. This captures the presence of several apparent rates for an activity in a component. In this type of functional dependency, the rate value expressed as a function of the current component state is still a positive real number and can never be zero. However, this may not always be the case if the functional dependency involves two or more components. For example, a functional dependency between two components means that the behaviour of one component depends on the current state of the other one. This implies that either the activity to be performed by the first component and/or its rate value will be determined by the current state of the second component. The rate value may then be any non-negative real number of $\mathbb{R}^+$ including zero, particularly when the choice of the activity to be performed is done according to the state of another component.

The introduction of functional dependencies in PEPA requires us to relax the constraint on the definition domain of an activity rate [15]. Thus, the set of activities $\mathcal{Act}$ is now defined as $\mathcal{Act} \subseteq \mathcal{A} \times \mathbb{R}^*$ where $\mathbb{R}^*$ is the set of positive real numbers defined as follows:

$$\mathbb{R}^* = \{r | r \geq 0; r \in \mathbb{R}\} \cup \{\top\}$$

For more details about the impact of functional dependencies on PEPA models and the aggregation technique see [15].

Example. Consider again the system of the previous example. If we opt for functional rates, the buffer may then be modelled differently:

$Buffer_0 \stackrel{def}{=} (in, \lambda).Buffer_1$

$Buffer_1 \stackrel{def}{=} (service, f \times p).Buffer_0 + (in, \lambda).Buffer_2$

$Buffer_2 \stackrel{def}{=} (service, f \times p).Buffer_1 + (service, f \times p).Buffer_3$

$Buffer_3 \stackrel{def}{=} (service, f \times p).Buffer_1$

where f is a function of the state i, $i = 0, \ldots, 3$ of component $Buffer$ such that:

$$f(i) = \begin{cases} s & \text{if } i = 1 \\ t & \text{if } i = 2, 3 \end{cases}$$

and p is a probability function defined as:

$$p(i) = \begin{cases} 1 & \text{if } i = 1, 3 \\ \frac{1}{2} & \text{if } i = 2 \end{cases}$$

Note that the definition of the $Server$, and the equation defining the complete system behaviour, remain unchanged:

$$System \stackrel{def}{=} Buffer_0 \underset{\{service\}}{\bowtie} Server$$

This example shows that the introduction of functional rates in PEPA models allows us to avoid having different apparent rates for an activity within a single component. Whereas in the first version of the model, activity $service$ has two apparent rates (s and t), the same activity has only one apparent rate (f) when using functional rates.

The association of an apparent rate with each action type within a single component leads, as we will show in Sect. 4, to a simplified tensor representation of the generator matrix associated with a PEPA model.

4   Tensor Representation

In this section we establish how to represent the infinitesimal generator matrix corresponding to a PEPA model as a sum of tensor products, analogous to the representation of a SAN. We proceed in three steps. In the first, we consider only the non-shared activities which do not belong to any cooperation set, representing the independent aspect of a component's behaviour. For each component we capture its local transitions in an appropriate matrix. In the second step, we consider the activities which belong to at least one cooperation set. This will allow us to take into account, in our tensorial representation, the interactions between the components. We represent each type of interaction by the tensor product of matrices capturing each component's capacity to participate in a shared activity. Using the obtained results, we finally show how to represent the global generator matrix of a complete model.

4.1   Non-interacting PEPA Components

We define a non-interacting component as a component for which at least one of its activities is a non-shared activity, or for which all its activities are cooperating activities, but there exists at least one non-shared activity in the model in which this component does not participate.

With each non-interacting component C_i, $i \in \{1, .., N\}$, we associate a generator matrix R_i of size $n_i \times n_i$ with $n_i = |ds(C_i)|$. If this component has at least one non-shared activity, the elements of its matrix are the rates of its individual activities. Otherwise, the matrix associated with this component is a null matrix. In both cases, the resulting matrix describes the local transitions of the component.

Now consider a PEPA model $M \stackrel{def}{=} C_1 \parallel C_2 \parallel \ldots \parallel C_N$ and assume that C_1, C_2, ..., C_N are represented by infinitesimal generator matrices R_1, R_2, ..., R_N respectively. Then any state of M can be represented as $(C_{1,j_1}, C_{2,j_2}, \ldots, C_{N,j_N})$ where $j_i \in \{1, 2, \ldots, n_i\}$ for $1 \leq i \leq N$. Moreover, the system of the N non-interacting components may be characterised by the infinitesimal generator matrix [17]

$$Q = \bigoplus_{k=1}^{N} R_k = \sum_{k=1}^{N} I_{n_1} \otimes \cdots \otimes I_{n_{k-1}} \otimes R_k \otimes I_{n_{k+1}} \otimes \cdots \otimes I_{n_N}$$
$$= \sum_{k=1}^{N} \bigotimes_{i=1}^{k-1} I_{n_i} \otimes R_k \otimes \bigotimes_{i=k+1}^{N} I_{n_i}$$

where I_d is the identity matrix of size d. $\oplus$ and $\otimes$ are the tensorial sum and product operators respectively [11].

4.2   Interacting PEPA Components

Most useful PEPA models are comprised of components which interact. To represent the interacting part of the component's behaviour, we associate with each action type α in $\mathcal{Z}$, the set of cooperating action types, a transition probability matrix $P_{i,\alpha}$. This matrix captures the capacity of component C_i to participate in the shared activity α. Thus, each element of this matrix represents the transition probability of component C_i with activity α with rate $r_\alpha(C_i)$. Note that if a component does not participate in a shared activity α, the matrix associated is an identity matrix.

In general, if a PEPA model is composed of N components, the interaction between these components can be expressed as follows:

$$\sum_{\alpha \in \mathcal{Z}} r_\alpha \bigotimes_{i=1}^{N} P_{i,\alpha}$$

where r_α is the minimum of the functional rates of action type α over all components C_i, $i = 1 \dots N$:

$$r_\alpha = min(r_\alpha(C_1), r_\alpha(C_2), \dots, r_\alpha(C_N))$$

4.3   Global Generator Matrix Representation

Now consider a PEPA model composed of interacting and non-interacting components. The corresponding generator matrix may be represented using Kronecker algebra as stated in Definition 1.

Definition 1. *The generator matrix Q of the Markov chain associated with a PEPA model is*

$$Q = \bigoplus_{i=1}^{N} R_i + \sum_{\alpha \in \mathcal{Z}} r_\alpha \left(\bigotimes_{i=1}^{N} P_{i,\alpha} - \bigotimes_{i=1}^{N} \overline{P}_{i,\alpha} \right) \tag{4.1}$$

where

- *N is the total number of components in the PEPA model and $\mathcal{Z}$ the set of cooperating action types, determined syntactically.*
- *r_α is the minimum of the functional rates of action type α over all components C_i, $i = 1 \dots N$.*
- *R_i is the transition matrix of component C_i relating solely to its individual activities.*
- *$P_{i,\alpha}$ is the probability transition matrix of component C_i due to activity of type α. Its elements' values are between 0 and 1.*
- *$\overline{P}_{i,\alpha}$ is a matrix representing the normalization associated with the shared activity α in component C_i.*

Unlike the local transition matrices R_i, the cooperation matrices $P_{i,\alpha}$ are not generators. So we need to introduce diagonal corrector matrices $\overline{P}_{i,\alpha}$ to normalize the cooperation matrices, i.e. ensure that row sums are zero. This is shown in (4.1).

In order to apply the equation above we must place a restriction on the use of types within cooperation sets, to ensure that each action type uniquely defines a synchronisation event. To see the need for this restriction, consider the model $M \stackrel{def}{=} (V \underset{\{\alpha\}}{\bowtie} S) \parallel (T \underset{\{\alpha\}}{\bowtie} U)$. Assuming that each component enables α with just one apparent rate $r_\alpha(V)$, $r_\alpha(S)$, etc., applying the equation above, we write the generator matrix of this model as follows:

$$Q = \bigoplus_{i=1}^{4} R_i + min(r_\alpha(V), r_\alpha(S), r_\alpha(T), r_\alpha(U)) \, (P_{V,\alpha} \otimes P_{S,\alpha} \otimes P_{T,\alpha} \otimes P_{U,\alpha}$$

$$- \overline{P}_{V,\alpha} \otimes \overline{P}_{S,\alpha} \otimes \overline{P}_{T,\alpha} \otimes \overline{P}_{U,\alpha})$$

It is clear that in this representation all the components are forced to make the *same* α cooperation. However, applying the semantics, there are two potential shared α activities: one involving V and S and one involving T and U. These can proceed concurrently with each other.

Thus we preprocess the model: when the same action type appears in distinct cooperation sets we rename the action type in the appropriate components and cooperation sets so that they are distinguished in $\mathcal{Z}$. For example, in the model above, we might distinguish α_ℓ (affecting V and S) and α_r (affecting T and U) and rename all α activities in V, S, T and U appropriately.

4.4 Example

Consider the two place buffer and the server described in Sect. 2. In the following we show how we construct the tensor expression for the global generator matrix of the corresponding model.

The model has two components, each component has two action types in its complete action type set: *in* and *service* for *Buffer* and *service* and *depart* for *Server*. The type *service* is the only element of $\mathcal{Z}$, the set of cooperating action types; the other action types being local to their respective components. Firstly we construct the matrices representing these local activities as follows:

$$R_{Buffer} = \begin{pmatrix} -\lambda & \lambda & 0 & 0 \\ 0 & -\lambda & \lambda & 0 \\ 0 & 0 & 0 & 0 \\ 0 & 0 & 0 & 0 \end{pmatrix} \qquad R_{Server} = \begin{pmatrix} 0 & 0 \\ d & -d \end{pmatrix}$$

When we come to represent the cooperations we consider each action type in the cooperating set. In our case, this set is composed of only one action type $\alpha = service$. Component *Buffer* participates to this activity with rate $r_\alpha(Buffer) = f$ whereas component *Server* participates to this activity with rate $r_\alpha(Server) = \top$. According to the semantic of PEPA, the resulting rate r_α of the shared activity is $r_\alpha = min(f, \top) = f$. Thus the *Buffer* component's contribution and the *Server* component's contribution to the cooperation are expressed by respectively:

$$P_{Buffer,\alpha} = \begin{pmatrix} 0 & 0 & 0 & 0 \\ 1 & 0 & 0 & 0 \\ 0 & \frac{1}{2} & 0 & \frac{1}{2} \\ 0 & 1 & 0 & 0 \end{pmatrix} \qquad \text{and} \qquad P_{Server,\alpha} = \begin{pmatrix} 0 & 1 \\ 0 & 0 \end{pmatrix}$$

The corresponding normalising matrix pairs are straightforward to construct:

$$\overline{P}_{Buffer,\alpha} = \begin{pmatrix} 0 & 0 & 0 & 0 \\ 0 & 1 & 0 & 0 \\ 0 & 0 & 1 & 0 \\ 0 & 0 & 0 & 1 \end{pmatrix} \qquad \text{and} \qquad \overline{P}_{Server,\alpha} = \begin{pmatrix} 1 & 0 \\ 0 & 0 \end{pmatrix}$$

Thus the complete expression becomes

$$Q = \begin{pmatrix} -\lambda & \lambda & 0 & 0 \\ 0 & -\lambda & \lambda & 0 \\ 0 & 0 & 0 & 0 \\ 0 & 0 & 0 & 0 \end{pmatrix} \oplus \begin{pmatrix} 0 & 0 \\ d & -d \end{pmatrix}$$

$$+ f \times \left[\begin{pmatrix} 0 & 0 & 0 & 0 \\ 1 & 0 & 0 & 0 \\ 0 & \frac{1}{2} & 0 & \frac{1}{2} \\ 0 & 1 & 0 & 0 \end{pmatrix} \otimes \begin{pmatrix} 0 & 1 \\ 0 & 0 \end{pmatrix} - \begin{pmatrix} 0 & 0 & 0 & 0 \\ 0 & 1 & 0 & 0 \\ 0 & 0 & 1 & 0 \\ 0 & 0 & 0 & 1 \end{pmatrix} \otimes \begin{pmatrix} 1 & 0 \\ 0 & 0 \end{pmatrix} \right]$$

leading to the complete generator matrix:

$$Q = \begin{pmatrix} -\lambda & 0 & \lambda & 0 & 0 & 0 & 0 & 0 \\ d & -(d+\lambda) & 0 & \lambda & 0 & 0 & 0 & 0 \\ 0 & s & -(s+\lambda) & 0 & \lambda & 0 & 0 & 0 \\ 0 & 0 & d & -(d+\lambda) & 0 & \lambda & 0 & 0 \\ 0 & 0 & 0 & s & -2s & 0 & 0 & s \\ 0 & 0 & 0 & 0 & d & -d & 0 & 0 \\ 0 & 0 & 0 & t & 0 & 0 & -t & 0 \\ 0 & 0 & 0 & 0 & 0 & 0 & d & -d \end{pmatrix}$$

Let us consider a modification of the model in which the *Server*, instead of being passive with respect to *service* has a local rate x, such that $s < x < t$. Then the construction of the tensor expression proceeds in exactly the same way except that when we come to compute the resulting rate we obtain is $r_\alpha = min(f, x)$ which value depends on the current state of component *Buffer*. According to this, the generator matrix is:

$$Q = \begin{pmatrix} -\lambda & 0 & \lambda & 0 & 0 & 0 & 0 & 0 \\ d & -(d+\lambda) & 0 & \lambda & 0 & 0 & 0 & 0 \\ 0 & s & -(s+\lambda) & 0 & \lambda & 0 & 0 & 0 \\ 0 & 0 & d & -(d+\lambda) & 0 & \lambda & 0 & 0 \\ 0 & 0 & 0 & \frac{x}{2} & -x & 0 & 0 & \frac{x}{2} \\ 0 & 0 & 0 & 0 & d & -d & 0 & 0 \\ 0 & 0 & 0 & x & 0 & 0 & -x & 0 \\ 0 & 0 & 0 & 0 & 0 & 0 & d & -d \end{pmatrix}$$

4.5 Validity of the Kronecker Expression

In this section we prove the validity of the tensor expression Q given in (4.1), i.e. we show that for all reachable states the tensor expression gives us the same transition rates as the generator matrix Q^* derived from the semantics of PEPA via the labelled transition system. First, we establish some notation and terminology.

A PEPA model is given by components $(C_i)_{i \in [1..N]}$, each with state space $S_i = ds(C_i)$ and a model equation $M \stackrel{def}{=} C_1 \bowtie_{L_1} \cdots \bowtie_{L_{N-1}} C_N$. Syntactic analysis readily identifies $\mathcal{Z}$, the set of cooperating actions. For each component C_i we can identify the cooperations it must participate in, a set we denote $\mathcal{Z}_i$, where $\mathcal{Z}_i = \mathcal{Z} \cap \mathcal{A}(C_i)$. Conversely, for each cooperating action type α, we denote by $\mathcal{Z}(\alpha)$ the set of components which participate in α typed activities.

For the same model we may have two views of the global state space. The first is $ds(M)$, the derivative set of M generated by the operational semantics via the labelled transition system. The second is $S = \prod_{i=1}^{N} S_i$, the product state space, generated directly from the derivative sets of the components. In general, the constraints placed on the model by cooperation will mean that $ds(M) \subset S$, i.e. S will contain unreachable states.

We suppose that every local state space S_i is ordered—simply take the order generated by the breadth-first search carried out in the PEPA Workbench to build the labelled transition system. In the following we assume that both $ds(M)$ and S are ordered lexicographically according to the ordering within component state spaces and the vector representation of the state space.

We will write $\mathbf{C}$ to denote a vector $(C_1, \ldots, C_N)$, and $\mathbf{C}[C_i := C_i']$ to denote the vector obtained from $\mathbf{C}$ by substituting C_i' for C_i. We denote by Q^*, the *original transition matrix* of M defined as follows:

- For all $\mathbf{C}, \mathbf{C}' \in ds(M)$ such that $\mathbf{C} \neq \mathbf{C}'$, $Q^*(\mathbf{C}, \mathbf{C}')$ is the transition rate as usually defined for PEPA, the sum of activity rates on arcs linking $\mathbf{C}$ and $\mathbf{C}'$ in the derivation graph:

$$Q^*(\mathbf{C}, \mathbf{C}') = \sum_{\mathbf{C} \xrightarrow{(a,r)} \mathbf{C}'} r$$

The set of all transitions(activities) can be partitioned into individual and shared transitions, shared transitions can be further partitioned by action type. Thus the off-diagonal elements of Q^* can be expressed as a sum of matrices as follows:

$$Q^*(\mathbf{C}, \mathbf{C}') = \sum_{i=1}^{N} Q_i^*(\mathbf{C}, \mathbf{C}') + \sum_{\alpha \in \mathcal{Z}} Q_\alpha^*(\mathbf{C}, \mathbf{C}') \qquad \text{where:}$$

$$Q_i^*(\mathbf{C}, \mathbf{C}') = \sum_{\mathbf{C} \xrightarrow{(a,r)} \mathbf{C}'} r \qquad \text{where } \mathbf{C}' = \mathbf{C}[C_i := C_i'] \text{ and } \alpha \notin \mathcal{Z}$$

$$Q_\alpha^*(\mathbf{C}, \mathbf{C}') = q(\mathbf{C}, \mathbf{C}', \alpha) \qquad \text{where } \alpha \in \mathcal{Z}$$

- For all $\mathbf{C} \in ds(M)$, $Q^*(\mathbf{C}, \mathbf{C})$ is calculated such that the sum of the elements in a row of Q^* will be zero.

Theorem 1. *Consider a well-defined PEPA model with N components interacting via a set of action types $\mathcal{Z}$ as defined by a model equation*

$$M \stackrel{\text{def}}{=} C_1 \underset{L_1}{\bowtie} \cdots \underset{L_{N-1}}{\bowtie} C_N$$

Then the original transition matrix Q^* is such that, for all reachable states, $\mathbf{C}, \mathbf{C}' \in ds(M)$, $Q^*(\mathbf{C}, \mathbf{C}') = Q(\mathbf{C}, \mathbf{C}')$ where Q is defined as

$$Q = \bigoplus_{i=1}^{N} R_i + \sum_{\alpha \in Z} min(r_\alpha(C_1), \ldots, r_\alpha(C_N)) \left(\bigotimes_{i=1}^{N} P_{i,\alpha} - \bigotimes_{i=1}^{N} \overline{P}_{i,\alpha} \right)$$

Furthermore, for $\mathbf{C} \in ds(M)$ and $\mathbf{C}' \notin ds(M)$, $Q(\mathbf{C}, \mathbf{C}') = 0$, i.e. there are no transitions from reachable states to unreachable ones represented in the tensor expression.

Proof. We can re-express Q as follows:

$$Q = \sum_{i=1}^{N} G_i + \sum_{\alpha \in Z} G_\alpha - \sum_{\alpha \in Z} G_{\alpha,n}$$

where

$$G_i = \bigotimes_{j=1}^{i-1} I_{n_j} \otimes R_i \otimes \bigotimes_{j=i+1}^{N} I_{j_j} \tag{4.2}$$

$$G_\alpha = r_\alpha \times \bigotimes_{i=1}^{N} P_{i,\alpha} \tag{4.3}$$

$$G_{\alpha,n} = r_\alpha \times \bigotimes_{i=1}^{N} \overline{P}_{i,\alpha} \tag{4.4}$$

and $r_\alpha = min(r_\alpha(C_1), \ldots, r_\alpha(C_N))$.

First, we consider the non-diagonal elements of the matrices. We will find it convenient to use kronecker functions:

$$\delta(x, y) = \begin{cases} 1 & \text{if } x = y \\ 0 & \text{if } x \neq y \end{cases}$$

Individual transitions. From above,

$$G_i = \bigotimes_{j=1}^{i-1} I_{n_j} \otimes R_i \otimes \bigotimes_{j=i+1}^{N} I_{n_j}$$

thus

$$G_i(\mathbf{C}, \mathbf{C}') = R_i(C_i, C_i') \times \prod_{\substack{k=1 \\ k \neq i}}^{N} \delta(C_k, C_k')$$

By the definitions of R_i and Q_i^*, it follows that, for all $\mathbf{C} \in ds(M)$, for all $i \in \{1, \ldots, N\}$,

$$G_i(\mathbf{C}, \mathbf{C}') = \begin{cases} Q_i^*(C_i, C_i') & \text{if } \mathbf{C}' = \mathbf{C}[C_i := C_i'] \\ 0 & \text{otherwise} \end{cases}$$

Clearly, $\mathbf{C}' = \mathbf{C}[C_i, C_i']$ implies that $\mathbf{C}'$ is reachable. Thus it follows that for $\mathbf{C}, \mathbf{C}' \in ds(M)$, for local transitions, i.e. those involving only one component C_i, the off-diagonal elements of Q_i^* and G_i are identical. Moreover, for $\mathbf{C} \in ds(M)$ and $\mathbf{C}' \notin ds(M)$, $G_i(\mathbf{C}, \mathbf{C}') = 0$.

Cooperating transitions. From above,

$$G_\alpha = r_\alpha \times \bigotimes_{i=1}^{N} P_{i,\alpha} \tag{4.5}$$

Thus

$$G_\alpha(\mathbf{C}, \mathbf{C}') = r_\alpha \times \prod_{k=1}^{N} P_{i,\alpha}(C_i, C_i')$$

Since a component i does not participate in activities of type α if $i \notin Z(\alpha)$, we can rewrite this as:

$$G_\alpha(\mathbf{C}, \mathbf{C}') = r_\alpha \times \prod_{i \in Z(\alpha)} P_{i,\alpha}(C_i, C_i') \times \prod_{i \notin Z(\alpha)} \delta(C_i, C_i')$$

Recall that $Q_\alpha^*(\mathbf{C}, \mathbf{C}') = q(\mathbf{C}, \mathbf{C}', \alpha)$, where $\alpha \in Z$. If we consider the semantic rule governing cooperation we can see that this transition rate consists of the minimal apparent rate of the participating components multiplied by the conditional probability that $\mathbf{C}'$ is the derivative resulting from the transition. This conditional probability is the product of the conditional probabilities in each component, since we assume that each component chooses between instances of α independently. For component C_i, this conditional probability is expressed as

$$p(C_i, C_i', \alpha) = \begin{cases} \frac{q(C_i, C_i', \alpha)}{q(C_i, \alpha)} & \text{if } i \in Z(\alpha) \\ 1 & \text{if } i \notin Z(\alpha) \text{ and } C_i = C_i' \\ 0 & \text{otherwise} \end{cases}$$

Thus, for all $\mathbf{C} \in ds(M)$

$$G_\alpha(\mathbf{C}, \mathbf{C}') = \begin{cases} r_\alpha \times p(\mathbf{C}, \mathbf{C}', \alpha) & \text{if } \mathbf{C}' \in ds(M) \\ 0 & \text{otherwise} \end{cases}$$

and so, for $\mathbf{C}, \mathbf{C}' \in ds(M)$,

$$G_\alpha(\mathbf{C}, \mathbf{C}') = r_\alpha \times p(\mathbf{C}, \mathbf{C}', \alpha)$$

It follows that for all $\mathbf{C} \in ds(M)$,

$$G_\alpha(\mathbf{C}, \mathbf{C}') = \begin{cases} Q_\alpha^*(\mathbf{C}, \mathbf{C}') & \text{if } \mathbf{C}' \in ds(M) \text{ and } \mathbf{C} \xrightarrow{(\alpha, r)} \mathbf{C}' \\ 0 & \text{otherwise} \end{cases}$$

In particular, for $\mathbf{C} \in ds(M)$ and $\mathbf{C}' \notin ds(M)$, $G_\alpha(\mathbf{C}, \mathbf{C}') = 0$ so there are no cooperating transitions into unreachable states from reachable ones.

Diagonal Elements. Finally, we consider only reachable states $\mathbf{C} \in ds(M)$ and show that $Q^*(\mathbf{C}, \mathbf{C}) = Q(\mathbf{C}, \mathbf{C})$. However, this follows immediately since by the previous arguments the off-diagonal elements of rows corresponding to $\mathbf{C}$ in Q^* and Q are in one-to-one correspondence, and furthermore, in each matrix the diagonal elements are chosen to normalise the matrix. For Q, G_i is already a generator whereas we have introduced $G_{\alpha,n}$ to normalise G_α. Thus by construction it follows that for all reachable states $\mathbf{C} \in ds(M)$, $Q^*(\mathbf{C}, \mathbf{C}) = Q(\mathbf{C}, \mathbf{C})$, as required.

□

According to the tensorial form of the generator, we store at most E entries:

$$E = (1 + 2|\mathcal{Z}|) \sum_{i=1}^{N} S_i^2$$

4.6   Solution Techniques

The tensorial representation of the generator matrix corresponding to a SAN model was proposed in 1984 by Plateau [18]. Since then different solution techniques have been investigated and several of them have been adapted to the context of this compact representation.

The main solution techniques used are either iterative methods such as the power method and Gauss Seidel or projective methods such as the Arnoldi and the GMRES methods.

In [21], the problem of computation time has been addressed. It has been shown how the methods of Arnoldi and GMRES can be used to substantially reduce the number of iterations needed when compared with the power method. Moreover, several preconditioning strategies that may be used to speed the iteration process even further have been investigated.

The power method, Arnoldi and GMRES methods have been incorporated in the tool PEPS implemented by Plateau's team. For each method, versions both with and without matrix preconditionning have been implemented.

The tensorial representation we propose for PEPA models is very similar to the one developed by Plateau for the SAN formalism. Therefore the solution techniques adapted to the tensorial representation of SAN and the computational results such as those presented in [10] about the impact of the functionnal rates on the Descriptor-vector multiplications in SAN may be, without doubt, applied in the context of PEPA formalism.

5   Related Work

Kronecker algebra representations have been used for some time as a means to address the state space explosion problem arising in the numerical solution of Markov chains. As mentioned earlier, the pioneering work in this area was carried out by Plateau on Stochastic Automata Networks [18]. More recently,

Kronecker-based solution techniques have been developed for various Petri net based-formalisms, for example [8,4,5,6].

With their explicit compositional structure, SPAs would appear to be natural candidates for Kronecker representation; however, there is little previous work on this topic. In 1994 Buchholz proposed an SPA called *MPA*, for which the mapping to an underlying Markov process is only defined in terms of a tensor expression [3]. However, in MPA the interpretation of both basic actions and shared actions is quite different to that in PEPA, chosen specifically to facilitate the tensor representation and without a natural modelling interpretation. MPA has not been developed further. In this approach the usual labelled transition system semantics is avoided and so there was no need to show the validity of the tensor expression with respect to the standard Markov process generation procedure. A similar denotational approach to semantics, making use of tensor expressions, is developed in the work of Rettelbach and Siegle [19].

In [9] El-Rayes presents an extension of PEPA and an associated solution technique based on the Matrix-Geometric Method (MGM). Her language $PEPA_{ph}^{\infty}$ allows exponential durations to be replaced by phase type distributions. In her mapping to the underlying Markov process, these distributions are represented by Kronecker expressions within the block-structured matrices used for the MGM. This is distinct from the use of Kronecker expressions in this paper.

6   Conclusions

In this paper we have presented a mapping from a SPA formalism to a Kronecker representation. Ours is the first such mapping aimed at implementation and incorporation into a tool. The SPA we use is PEPA and the mapping is specific to that formalism due to the complex semantic rules defining synchronisation between PEPA components. Whilst other SPAs such as EMPA [2] and IMC [12] have apparently simpler rules of synchronisation, they include immediate actions which would complicate the mapping to a Kronecker representation.

Once the prototype implementation of this approach is incorporated into the PEPA Workbench [7], we aim to improve its efficiency. In particular we plan to investigate the multilevel approaches which have been employed with SPN to avoid the incorporation of unreachable states. We will also be interested in investigating techniques which exploit this Kronecker representation to solve the model efficiently and in comparing our approach with other compact representations, such as those based on BDDs [13].

References

1. Ajmone Marsan, A., Conte, G., Balbo, G.: A class of generalised stochastic Petri nets for the performance evaluation of multiprocessor systems. ACM Transactions on Computer Systems **2(2)** (1984) 93–122
2. Bernardo, M., Gorrieri, R.: A Tutorial on EMPA: A theory of concurrent processes with nondeterminism, probabilities and time. Theoretical Computer Science. **201** (1998) 1–54

3. Buchholz, B.: Compositional analysis of a Markovian Process Algebra. Proc. of 2nd Process Algebra and Performance Modelling Workshop. In U. Herzog and M. Rettelbach Editors (1994)
4. Buchholz, P., Kemper, P.: Numerical analysis of stochastic marked graphs. Proc. of Int. Workshop on Petri Nets and Performance Models. IEEE-Computer Society Press. Durham, NC (1995) 32–41
5. Campos, J., Donatelli, S., Silva, M.: Structured solution of stochastic DSSP systems. Proc. of Int. Workshop on Petri Nets and Performance Models. IEEE-Computer Society Press. St Malo, France (1997) 91–100
6. Ciardo, G., Miner, A.S.: A data structure for the efficient Kronecker solution of GSPNs. In P. Buchholz editor, Proc. of the 8th International Workshop on Petri Nets and Performance Models (PNPM'99) Zaragoza, Spain, (1999) 22–31
7. Clark, G., Gilmore, S., Hillston, J., Thomas, N.: Experiences with the PEPA performance modelling tools. IEE Software. **146(1)** (1999) 11–19
8. Donatelli, S.: Superposed Generalised stochastic Petri nets: definition and efficient solution. Proc. of 15th Int. Conf. on Application and Theory of Petri Nets. In M. Silva Editor (1994)
9. El-Rayes, A.: Analysing performance of open queueing systems with stochastic process algebra. University of Birmingham (2000)
10. Fernandes, P., Plateau, B., Stewart, W.J.: Efficient vector-descriptor multiplications in stochastic automata networks. INRIA Report #2935. Anonymous ftp. ftp.inria.fr/INRIA/Publication/RR
11. Graham, A.: Kronecker products and matrix calculus with applications. Prentice Hall (1989)
12. Hermanns, H.: Interactive Markov Chains. PhD Thesis, Universität Erlangen-Nürnberg (1999)
13. Hermanns, H., Meyer-Kayser, J., Siegle, M.: Multi-terminal binary decision diagrams to represent and analyse continuous time Markov chains. Proc. 3rd Int. Workshop on Numerical Solution of Markov Chain, Zaragoza, Spain, (1999) 188–207
14. Hillston, J.: A Compositional approach to performance modelling. PhD Thesis, The University of Edinburgh (1994)
15. Hillston, J., Kloul, L.: From SAN to PEPA: A technology transfer. Submitted.
16. Molloy, M.K.: Performance analysis using stochastic Petri nets. IEEE Transactions on Computers **31(9)** (1982) 913–917
17. Plateau, B.: On the stochastic structure of parallelism and synchronisation models for distributed algorithms. Proc. ACM Sigmetrics Conference on Measurement and Modelling of Computer Systems (1985)
18. Plateau, B.: De l'évolution du parallélisme et de la synchronisation. PhD Thesis, Université de Paris-Sud, Orsay (1984)
19. Rettelbach, M., Siegle, M.: Compositional minimal semantics for the stochastic process algebra TIPP. Proc. of 2nd Workshop on Process Algebra and Performance Modelling (1994)
20. Sanders, W.H., Meyer, J.F.: Reduced base model construction methods for stochastic activity networks. IEEE Journal on Selected Areas in Communications **9(1)** (1991) 25–36
21. Stewart, W.J., Atif, K., Plateau, B.: The numerical solution of stochastic automata networks. European Journal of Operations Research **86(3)** (1995) 503–525

Reward Based Congruences:
Can We Aggregate More?

Marco Bernardo[1] and Mario Bravetti[2]

[1] Università di Torino, Dipartimento di Informatica
Corso Svizzera 185, 10149 Torino, Italy
bernardo@di.unito.it
[2] Università di Bologna, Dipartimento di Scienze dell'Informazione
Mura Anteo Zamboni 7, 40127 Bologna, Italy
bravetti@cs.unibo.it

Abstract. In this paper we extend a performance measure sensitive Markovian bisimulation congruence based on yield and bonus rewards that has been previously defined in the literature, in order to aggregate more states and transitions while preserving compositionality and the values of the performance measures. The extension is twofold. First, we show how to define a performance measure sensitive Markovian bisimulation congruence that aggregates bonus rewards besides yield rewards. This is achieved by taking into account in the aggregation process the conditional execution probabilities of the transitions to which the bonus rewards are attached. Second, we show how to define a performance measure sensitive Markovian bisimulation congruence that allows yield rewards and bonus rewards to be used interchangeably up to suitable correcting factors, aiming at the introduction of a normal form for rewards. We demonstrate that this is possible in the continuous time case, while it is not possible in the discrete time case because compositionality is lost.

1 Introduction

In the past five years the problem of specifying performance measures has been addressed in the field of Markovian concurrent systems. Following [10], in [6, 7,3] the performance measures are characterized through atomic rewards to be suitably attached to the states and the transitions of the Markov chains (MCs for short) associated with the Markovian process algebraic specifications of the systems. While in [6,7] the states to which certain rewards have to be attached are singled out by means of temporal logic formulas to be model checked, in [3] rewards are directly specified within the actions occurring in the specifications and are then trasferred to the MCs during their construction. In [13], instead, temporal reward formulas are introduced, which are able to express accumulated atomic rewards over sequences of states and allow performance measures to be evaluated on MCs through techniques for computing long run averages. Finally, in [1] a temporal logic is used to directly specify performance measures

L. de Alfaro and S. Gilmore (Eds.): PAPM-PROBMIV 2001, LNCS 2165, pp. 136–151, 2001.

for Markovian transition systems, where the evaluation of such performance measures is conducted via model checking.

Among the approaches mentioned above, in this paper we concentrate on that of [3]. Its distinguishing feature is that of being deeply rooted in the Markovian process algebraic formalism. This has allowed us to define a performance measure sensitive congruence in the bisimulation style inspired by [11], which permits to compositionally manipulate system specifications without altering the values of their performance measures. We recall that taking performance measures into account when e.g. compositionally minimizing the state space of a Markovian process algebraic specification is important. If for example the equivalence used in the minimization process gives rise to the merging of two states whose associated rewards are different, we come to the undesirable situation in which the original model and the minimized model result in two different values for the same performance measure.

Following [10], the reward based Markovian bisimulation equivalence of [3] considers two types of rewards: yield rewards, which are accumulated while staying in the states, and bonus rewards, which are instantaneously gained when executing the transitions. Such an equivalence essentially aggregates yield rewards whenever possible, but does not manipulate bonus rewards at all.

The contribution of this paper is to extend the equivalence of [3] in order to aggregate as much as possible while retaining the bisimulation style, the congruence property, and the value of the performance measures. The extension is twofold. First, we show how to define a Markovian bisimulation congruence that aggregates bonus rewards as well. This is achieved by taking into account in the aggregation process the conditional execution probabilities of the transitions to which the bonus rewards are attached. Second, we show how to define a Markovian bisimulation congruence that allows yield rewards and bonus rewards to be used interchangeably up to suitable correcting factors, aiming at the introduction of a normal form for rewards. More precisely, we demonstrate that this is possible in the continuous time case, while it is not possible in the discrete time case because compositionality is lost.

Since the way in which bonus rewards can be aggregated is easy to find and the way in which yield and bonus rewards can be interchanged is known in the literature, this paper is especially concerned with investigating whether they respect compositionality or not. We also observe that, although such an investigation is conducted for a particular language (EMPA$_{gr_n}$ [3]), its results can be applied to every Markovian process algebra.

This paper is organized as follows. In Sect. 2 we recall the basic notions about reward structures. In Sect. 3 we recall EMPA$_{gr_n}$, a process algebra for the specification of continuous time and discrete time Markovian systems, and the related reward based Markovian bisimulation congruence. In Sect. 4 we present an improved reward based Markovian bisimulation equivalence that aggregates bonus rewards as well, we prove that it is a congruence, and we show a sound and complete axiomatization. In Sect. 5 we present a further improved reward based Markovian bisimulation equivalence that allows yield rewards and bonus rewards to be used interchangeably in the continuous time case, we prove that it is a congruence, and we show a sound and complete axiomatization. In Sect. 6 we report some concluding remarks.

2   Reward Structures

In the performance evaluation area, the technique of rewards is frequently used to specify and derive measures for system models whose underlying stochastic process is a MC. According to [10], a reward structure for a MC is composed of: a yield function $y_{i,j}(t)$, expressing the rate at which reward is accumulated at state i t time units after i was entered when the successor state is j, and a bonus function $b_{i,j}(t)$, expressing the reward awarded upon exit from state i and subsequent entry into state j given that the holding time in state i was t time units. Since the generality of this structure is difficult to fully exploit due to the complexity of the resulting solution, the analysis is usually simplified by considering yield functions that do not depend on the time nor the successor state, as well as bonus functions that do not depend on the holding time of the previously occupied state: $y_{i,j}(t) = y_i$ and $b_{i,j}(t) = b_{i,j}$.

Several performance measures can be calculated by exploiting rewards. According to the classifications proposed in [12,9], we have instant-of-time measures, expressing the gain received at a particular time instant, and interval-of-time (or cumulative) measures, expressing the overall gain received over some time interval. Both kinds of measures can refer to stationary or transient state. In the following, we concentrate on instant-of-time performance measures.

In the stationary case, instant-of-time performance measures quantify the long run gain received per unit of time. Given yield rewards y_i and bonus rewards $b_{i,j}$ for a certain MC, the corresponding stationary performance measure is computed as:

$$\sum_i y_i \cdot \pi_i + \sum_i \sum_j b_{i,j} \cdot \phi_{i,j} \tag{1}$$

where π_i is the stationary probability of state i and $\phi_{i,j}$ is the stationary frequency with which the transition from state i to state j is traversed. $\phi_{i,j}$ is given by the stationary frequency with which state i is entered (i.e. the ratio of its stationary probability to its average holding time) multiplied by the probability with which the transition from state i to state j is traversed given that the current state is i. In the case of a continuous time MC (CTMC) we have $\phi_{i,j} = \pi_i \cdot q_{i,j}$ with $q_{i,j}$ being the rate of the transition from i to j, while in the case of a discrete time MC (DTMC) we have $\phi_{i,j} = \pi_i \cdot p_{i,j}$ with $p_{i,j}$ being the probability of the transition from i to j.

In the transient case, instant-of-time performance measures quantify the gain received at a specific time instant. Given yield rewards y_i and bonus rewards $b_{i,j}$ for a certain MC, the corresponding transient state performance measure is computed as:

$$\sum_i y_i \cdot \pi_i(t) + \sum_i \sum_j b_{i,j} \cdot \phi_{i,j}(t) \tag{2}$$

where $\pi_i(t)$ is the transient probability of being in state i at time t and $\phi_{i,j}(t)$ is the transient frequency with which the transition from state i to state j is traversed at time t, which is computed in the same way as $\phi_{i,j}$ with $\pi_i(t)$ in place of π_i.

3 An Overview of EMPA$_{gr_n}$

EMPA$_{gr_n}$ [3] is a family of extended Markovian process algebras with generative-reactive synchronizations [5] whose actions are enriched in order to accommodate the specification of $n \in \mathbf{N}$ different performance measures simultaneously. This is achieved by associating with every action a pair composed of a yield reward and a bonus reward for each performance measure. The number n is called the order. For the sake of simplicity, without loss of generality in this paper we deal with order 1 only.

3.1 Syntax and Informal Semantics

The main ingredients of our calculus are the actions, each composed of a type, a rate, and a sequence of n pairs of yield and bonus rewards, and the algebraic operators. As far as actions are concerned, based on their rates they are classified into exponentially timed (rate $\lambda \in \mathbf{R}_+$ representing the parameter of the corresponding exponentially distributed duration), immediate (rate $\infty_{l,w}$ to denote a zero duration, with $l \in \mathbf{N}_+$ being a priority level and $w \in \mathbf{R}_+$ being a weight associated with the action choice), and passive (rate $*_{l,w}$ to denote an unspecified duration with priority level l and weight w associated with the action choice). Moreover, based on their types, actions are classified into observable and invisible depending on whether they are different or equal to τ, as usual.

Definition 1. *Let $AType$ be the set of action types including the invisible type τ, $ARate = \mathbf{R}_+ \cup \{\infty_{l,w} \mid l \in \mathbf{N}_+ \wedge w \in \mathbf{R}_+\} \cup \{*_{l,w} \mid l \in \mathbf{N}_+ \wedge w \in \mathbf{R}_+\}$ be the set of action rates, $ARew = \mathbf{R} \cup \{*\}$ be the set of action rewards. We use a to range over $AType$, $\tilde{\lambda}$ to range over $ARate$, λ to range over exponentially timed rates, $\bar{\lambda}$ to range over active (i.e. nonpassive) rates, $\tilde{y}$ to range over yield rewards (y if not $*$), and $\tilde{b}$ to range over bonus rewards (b if not $*$). The set of actions of order 1 is defined by*

$$Act_1 = \{<a, \tilde{\lambda}, (\tilde{y}, \tilde{b})> \in AType \times ARate \times (ARew \times ARew) \mid$$
$$(\tilde{\lambda} \in \{*_{l,w} \mid l \in \mathbf{N}_+ \wedge w \in \mathbf{R}_+\} \wedge \tilde{y} = \tilde{b} = *) \vee$$
$$(\tilde{\lambda} \in \mathbf{R}_+ \cup \{\infty_{l,w} \mid l \in \mathbf{N}_+ \wedge w \in \mathbf{R}_+\} \wedge \tilde{y}, \tilde{b} \in \mathbf{R})\} \quad \blacksquare$$

Definition 2. *Let $Const$ be a set of constants ranged over by A and let $ATRFun = \{\varphi : AType \longrightarrow AType \mid \varphi^{-1}(\tau) = \{\tau\}\}$ be a set of action type relabeling functions ranged over by φ. The set $\mathcal{L}_1$ of process terms of EMPA$_{gr_1}$ is generated by the following syntax*

$$E ::= \underline{0} \mid <a, \tilde{\lambda}, (\tilde{y}_1, \tilde{b}_1)>.E \mid E/L \mid E[\varphi] \mid E + E \mid E \|_S E \mid A$$
where $L, S \subseteq AType - \{\tau\}$. $\quad \blacksquare$

The null term "$\underline{0}$" is the term that cannot execute any action.

The action prefix operator "$<a, \tilde{\lambda}, (\tilde{y}, \tilde{b})>._$" denotes the sequential composition of an action and a term. Term $<a, \tilde{\lambda}, (\tilde{y}, \tilde{b})>.E$ can execute an action with type a and rate $\tilde{\lambda}$, thus making the corresponding state earn additional yield reward $\tilde{y}$ and the related transition gain bonus reward $\tilde{b}$, and then behaves as term E.

The functional abstraction operator "$_/L$" abstracts from the type of the actions. Term E/L behaves as term E except that the type a of each executed action is turned into τ whenever $a \in L$.

The functional relabeling operator "$_[\varphi]$" changes the type of the actions. Term $E[\varphi]$ behaves as term E except that the type a of each executed action becomes $\varphi(a)$.

The alternative composition operator "$_ + _$" expresses a choice between two terms. Term $E_1 + E_2$ behaves as either term E_1 or term E_2 depending on whether an action of E_1 or an action of E_2 is executed. The choice among several enabled exponentially timed actions is solved according to the race policy, i.e. the fastest action is executed (this implies that each action has an execution probability proportional to its rate). If immediate actions are enabled as well, they take precedence over exponentially timed ones and the choice among them is solved according to the preselection policy: the immediate actions having the highest priority level are singled out, then each of them is given an execution probability proportional to its weight. The choice among several enabled passive actions is instead solved according to the reactive preselection policy: for each action type, the passive actions having the highest priority level are singled out, then each of them is given an execution probability proportional to its weight. Therefore, the choice among passive actions having different types and the choice between passive and active actions are nondeterministic.

The parallel composition operator "$_ \|_S _$" expresses the concurrent execution of two terms. Term $E_1 \|_S E_2$ asynchronously executes actions of E_1 or E_2 not belonging to S and synchronously executes actions of E_1 and E_2 belonging to S according to the two following synchronization disciplines. The synchronization discipline on action types establishes that two actions can synchronize if and only if they have the same observable type in S, which becomes the resulting type. Following the terminology of [8], the synchronization discipline on action rates is the generative-reactive mechanism, which establishes that two actions can synchronize if and only if at least one of them is passive (behaves reactively). In case of synchronization of an active action a having rate $\tilde{\lambda}$ executed by E_1 (E_2) with a passive action a having rate $*_{l,w}$ executed by E_2 (E_1), the resulting active action a has rate/weight given by the original rate/weight multiplied by the probability that E_2 (E_1) chooses the passive action at hand among its passive actions of type a. Instead, in case of synchronization of two passive actions a having rate $*_{l_1,w_1}$ and $*_{l_2,w_2}$ executed by E_1 and E_2, respectively, the resulting passive action of type a has priority level given by the maximum l_{max} between l_1 and l_2 and weight given by the probability that E_1 and E_2 independently choose the two actions, multiplied by a normalization factor given by the overall weight of the passive actions of type a executable by E_1 and E_2 at the priority level l_{max}. As far as rewards are concerned, since only the rewards of active actions are specified, in case of synchronization they are handled as follows. The yield rewards of an active action are treated exactly as the rate of that action, i.e. they are multiplied by the execution probabilities of the passive actions involved in the synchronization. Instead, the bonus rewards of an active action are just inherited, as multiplying them by the execution probabilities of the aforementioned passive actions would lead to an underestimation of the performance measures. The reason is that, in the calculation of the performance measures according to formulas (1) and (2), each bonus reward of a transition is multiplied by a factor that is proportional to the rate of the transition itself, hence multiplying the rates by the execution probabilities of passive actions is all we have to do. In the case of synchronization between two passive actions, the rewards of the resulting passive actions are still unspecified.

Finally, we assume the existence of a set of constant defining equations of the form $A \stackrel{\Delta}{=} E$. In order to guarantee the correctness of recursive definitions, as usual we restrict ourselves to the set $\mathcal{G}_1$ of terms that are closed and guarded w.r.t. Def_1.

3.2 Reward Master-Slaves Transition Systems

The semantic model of $\mathrm{EMPA}_{\mathrm{gr}_1}$ is a special kind of LTS called master-slaves transition system of order 1 (RMSTS_1 for short), whose transitions are labeled with elements of Act_1. Recalling that active actions play the role of the masters (they behave generatively) while passive actions play the role of the slaves (they behave reactively), each state of a RMSTS_1 has a single master bundle composed of all the transitions labeled with an active action and, for each action type a, a single slave bundle of type a composed of all the transitions labeled with a passive action of type a. Since the operational semantics for $\mathrm{EMPA}_{\mathrm{gr}_1}$ will be defined in such a way that lower priority active transitions are not pruned (in order to get a congruence) while lower priority passive transitions of a given type are, all the passive transitions belonging to the same slave bundle of a generated RMSTS_1 have the same priority level.

Definition 3. *A reward master-slaves transition system of order 1 ($RMSTS_1$) is a triple*

$$(S, AType, \longrightarrow)$$

where S is a set of states, $AType$ is a set of action types, and $\longrightarrow \in \mathcal{M}(S \times Act_1 \times S)$ is a multiset [1] of transitions such that for all $s \in S$ and $a \in AType$:

$$(s \xrightarrow{a, *_{l', w'}, (*, *)} s' \wedge s \xrightarrow{a, *_{l'', w''}, (*, *)} s'') \implies l' = l''$$

A rooted reward master-slaves transition system of order 1 ($RRMSTS_1$) is a quadruple

$$(S, AType, \longrightarrow, s_0)$$

where $(S, AType, \longrightarrow)$ is a $RMSTS_1$ and $s_0 \in S$ is the initial state. ∎

---

[1] We use "$\{\!|$" and "$|\!\}$" as brackets for multisets and $\mathcal{M}(S)$ ($\mathcal{P}(S)$) to denote the collection of multisets over (subsets of) S.

We point out that the transition relation is a multiset, not a set. This allows the multiplicity of identically labeled transitions to be taken into account, which is necessary from the stochastic point of view. As an example, if a state has two transitions both labeled with $<a, \lambda, (y, b)>$, using sets instead of multisets would reduce the two transitions into a single one with rate λ, thus erroneously altering the average sojourn time in the state.

Given a state, the choice among the bundles of transitions enabled in that state is nondeterministic. The choice of a transition within the master bundle is governed by the race policy if there are only exponentially timed transitions, the preselection policy if there are immediate transitions (which take precedence over exponentially timed transitions). The choice of a transition within a slave bundle of type a is governed by the preselection policy.

We observe that the passive actions are seen as incomplete actions that must synchronize with active actions of the same type of another system component in order to form a complete system. Therefore, a fully specified system is performance closed, in the sense that it gives rise to a fully probabilistic transition system that does not include slave bundles. If in such a transition system we keep for each state only the highest priority transitions, then we can easily derive a performance model in the form of a reward DTMC or CTMC, depending on whether only immediate transitions occur or not. We point out that, if only immediate transitions occur, each of them is assumed to take one time step, hence the underlying stochastic model naturally turns out to be a DTMC. Should exponentially timed and immediate transitions coexist (in different states), a CTMC is derived by eliminating the immediate transitions and the related source states and by suitably splitting the exponentially timed transitions entering the removed source states, in such a way that they are caused to reach the target states of the removed immediate transitions.

As far as the yield rewards are concerned, when constructing a reward MC from a $RRMSTS_1$ we proceed as follows. Whenever a state has several actions, be it due to an alternative composition operator or a parallel composition operator, we make the additivity assumption, i.e. we assume that the yield reward earned by the state is the sum of the yield rewards of its transitions. This assumption is consistent with the race inherent in the parallel composition operator and with the adoption of the race policy for the alternative composition operator, i.e. with viewing alternative actions as being in parallel execution, hence all contributing to the reward accumulation in the state.

3.3   Operational Semantics

The formal semantics for $EMPA_{gr_1}$ maps terms onto $RRMSTS_1$. We preliminarily provide the following shorthands to make the definition of the operational semantic rules easier.

Definition 4. *Given a $RMSTS_1$ $M = (S, AType, \longrightarrow)$, $s \in S$, and $a \in AType$, we denote by $L_a(s)$ the priority level of the slave transitions of type*

a executable at s ($L_a(s) = 0$ *if the slave bundle a of s is empty) and we de-*
note by $W_a(s)$ *the overall weight of the slave transitions of type a executable at s:*

$$W_a(s) = \sum \{ w \mid \exists s' \in S. s \xrightarrow{a, *_{L_a(s), w, (*, *)}} s' \}$$

Furthermore, we extend the real number multiplication to immediate rates as
follows:

$$\infty_{l,w} \cdot p = \infty_{l, w \cdot p} \qquad \blacksquare$$

The operational semantics for $EMPA_{gr_1}$ is the least $RMSTS_1$ $(\mathcal{G}_1, AType, \longrightarrow_1)$ satisfying the inference rules of Table 1, where in addition to the rules $(Ch1_l)$, $(Ch2_l)$, $(Pa1_l)$, $(Pa2_l)$, $(Sy1_l)$ referring to a move of the lefthand process E_1, we consider also the symmetrical rules $(Ch1_r)$, $(Ch2_r)$, $(Pa1_r)$, $(Pa2_r)$, $(Sy1_r)$ taking into account the moves of the righthand process E_2, obtained by exchanging the roles of terms E_1 and E_2. We consider the operational rules as generating a multiset of transitions (consistently with the definition of $RMSTS_1$), where a transition has arity m if and only if it can be derived in m possible ways from the operational rules.

Some explanations are now in order. First of all, the operational rules give rise to an interleaving semantics, which is made possible by the memoryless property of exponential distributions. The removal of lower priority passive transitions of the same type is carried out in rules $(Ch2_l)$ and $(Ch2_r)$ for the alternative composition operator and rules $(Pa1_l)$ and $(Pa1_r)$ for the parallel composition operator by using $L_a(E)$.

In the case of a synchronization, the evaluation of the rate of the resulting action is carried out by rules $(Sy1_l)$, $(Sy1_r)$, and $(Sy2)$ as follows. Whenever an active action synchronizes with a passive action of the same type, the rate of the resulting active action is evaluated in rules $(Sy1_l)$ and $(Sy1_r)$ by multiplying the rate of the active action by the probability of choosing the passive action. The yield reward of the active action undergoes the same treatment, while the bonus reward is just inherited. Whenever two passive actions of type a synchronize, instead, the priority level and the weight of the resulting passive action are computed as described by rule $(Sy2)$. In particular, the weight is computed by multiplying the probability p of independently choosing the two original actions by the normalization factor N, which is given by the overall weight of the passive transitions of type a with maximum priority level executable by E_1 and E_2, computed by using $W_a(E)$.

Definition 5. *The integrated semantics of* $E \in \mathcal{G}_1$ *is the* $RRMSTS_1$
$$\mathcal{I}_1[\![E]\!] = (\mathcal{G}_{1,E}, AType, \longrightarrow_{1,E}, E)$$
where $\mathcal{G}_{1,E}$ *is the set of terms reachable from* E *according to the* $RMSTS_1$ $(\mathcal{G}_1, AType, \longrightarrow_1)$ *and* $\longrightarrow_{1,E}$ *is the restriction of* $\longrightarrow_1$ *to transitions between terms in* $\mathcal{G}_{1,E}$. *We say that* $E \in \mathcal{G}_1$ *is performance closed if and only if* $\mathcal{I}_1[\![E]\!]$ *does not contain passive transitions. We denote by* $\mathcal{E}_1$ *the set of performance closed terms of* $\mathcal{G}_1$. $\qquad \blacksquare$

We conclude by recalling that from $\mathcal{I}_1[\![E]\!]$ two projected semantic models can be obtained by essentially dropping action rates or action types, respectively. Before applying such a transformation to $\mathcal{I}_1[\![E]\!]$, lower priority active transitions

Table 1. $EMPA_{gr_1}$ operational semantics

$$(Pr) \quad <a,\tilde{\lambda},(\tilde{y},\tilde{b})>.E \xrightarrow{a,\tilde{\lambda},(\tilde{y},\tilde{b})}_1 E$$

$$(Hi1) \quad \frac{E \xrightarrow{a,\tilde{\lambda},(\tilde{y},\tilde{b})}_1 E'}{E/L \xrightarrow{a,\tilde{\lambda},(\tilde{y},\tilde{b})}_1 E'/L} \quad a \notin L \qquad\qquad (Hi2) \quad \frac{E \xrightarrow{a,\tilde{\lambda},(\tilde{y},\tilde{b})}_1 E'}{E/L \xrightarrow{\tau,\tilde{\lambda},(\tilde{y},\tilde{b})}_1 E'/L} \quad a \in L$$

$$(Re) \quad \frac{E \xrightarrow{a,\tilde{\lambda},(\tilde{y},\tilde{b})}_1 E'}{E[\varphi] \xrightarrow{\varphi(a),\tilde{\lambda},(\tilde{y},\tilde{b})}_1 E'[\varphi]}$$

$$(Ch1_l) \quad \frac{E_1 \xrightarrow{a,\tilde{\lambda},(y,b)}_1 E_1'}{E_1 + E_2 \xrightarrow{a,\tilde{\lambda},(y,b)}_1 E_1'} \qquad\qquad (Ch2_l) \quad \frac{E_1 \xrightarrow{a,*_l,w,(*,*)}_1 E_1' \quad l \geq L_a(E_2)}{E_1 + E_2 \xrightarrow{a,*_l,w,(*,*)}_1 E_1'}$$

$$(Pa1_l) \quad \frac{E_1 \xrightarrow{a,\tilde{\lambda},(y,b)}_1 E_1'}{E_1 \parallel_S E_2 \xrightarrow{a,\tilde{\lambda},(y,b)}_1 E_1' \parallel_S E_2} \quad a \notin S$$

$$(Pa2_l) \quad \frac{E_1 \xrightarrow{a,*_l,w,(*,*)}_1 E_1' \quad l \geq L_a(E_2)}{E_1 \parallel_S E_2 \xrightarrow{a,*_l,w,(*,*)}_1 E_1' \parallel_S E_2} \quad a \notin S$$

$$(Sy1_l) \quad \frac{E_1 \xrightarrow{a,\tilde{\lambda},(y,b)}_1 E_1' \quad E_2 \xrightarrow{a,*_l,w,(*,*)}_1 E_2'}{E_1 \parallel_S E_2 \xrightarrow{a,\tilde{\lambda}\cdot\frac{w}{W_a(E_2)},(y\cdot\frac{w}{W_a(E_2)},b)}_1 E_1' \parallel_S E_2'} \quad a \in S$$

$$(Sy2) \quad \frac{E_1 \xrightarrow{a,*_{l_1,w_1},(*,*)}_1 E_1' \quad E_2 \xrightarrow{a,*_{l_2,w_2},(*,*)}_1 E_2'}{E_1 \parallel_S E_2 \xrightarrow{a,*_{\max(l_1,l_2),p\cdot N},(*,*)}_1 E_1' \parallel_S E_2'} \quad a \in S$$

$$\text{where:} \quad p = \frac{w_1}{W_a(E_1)} \cdot \frac{w_2}{W_a(E_2)} \qquad N = \begin{cases} W_a(E_1) + W_a(E_2) & \text{if } l_1 = l_2 \\ W_a(E_1) & \text{if } l_1 > l_2 \\ W_a(E_2) & \text{if } l_2 > l_1 \end{cases}$$

$$(Co) \quad \frac{E \xrightarrow{a,\tilde{\lambda},(\tilde{y},\tilde{b})}_1 E'}{A \xrightarrow{a,\tilde{\lambda},(\tilde{y},\tilde{b})}_1 E'} \quad A \stackrel{\Delta}{=} E$$

are pruned because E is no longer to be composed with other terms as it describes the whole system we are interested in. The functional semantics $\mathcal{F}_1[\![E]\!]$ is a standard LTS whose transitions are decorated with action types only. The Markovian semantics $\mathcal{M}_1[\![E]\!]$ is instead a reward CTMC or DTMC, as seen in Sect. 3.2, which is well defined only if E is performance closed.

3.4   Reward Based Markovian Bisimulation Equivalence

$EMPA_{gr_1}$ is equipped with a reward based Markovian bisimulation equivalence, which relates two systems having the same functional, probabilistic, prioritized and exponentially timed behavior, as well as the same performance measure values, by considering their ability to simulate each other behavior. To achieve this, the rates/weights of the transitions of the same type and priority level that leave the same state and reach states belonging to the same equivalence class are summed up, like in the exact aggregation for MCs known as ordinary lumping. The yield rewards labeling the transitions above are handled in the same way as the corresponding rates, because of the additivity assumption. The bonus rewards of the transitions above, instead, are not summed up, as this would result in an overestimation of the specified performance measures. The reason is that, in the calculation of the performance measures according to formulas (1) and (2), the bonus reward of a transition is multiplied by a factor that is proportional to the rate of the transition itself, hence summing rates up is all we have to do.

Definition 6. *We define function priority level* $PL : ARate \longrightarrow \mathbb{Z}$ *by:*
$$PL(*_{l,w}) = -l$$
$$PL(\lambda) = 0$$
$$PL(\infty_{l,w}) = l$$
and we extend the real number summation to rates of the same priority level and to unspecified rewards as follows:
$$*_{l,w_1} + *_{l,w_2} = *_{l,w_1+w_2}$$
$$\infty_{l,w_1} + \infty_{l,w_2} = \infty_{l,w_1+w_2}$$
$$* + * = *$$
We define partial function aggregated rate-yield $RY_1 : \mathcal{G}_1 \times AType \times \mathbb{Z} \times ARew \times \mathcal{P}(\mathcal{G}_1) \longrightarrow ARate \times ARew$ *by:*
$$RY_1(E, a, l, \tilde{b}, C) = (Rate_1(E, a, l, \tilde{b}, C), Yield_1(E, a, l, \tilde{b}, C))$$
where:
$$Rate_1(E, a, l, \tilde{b}, C) = \sum\{\!| \tilde{\lambda} \mid \exists \tilde{y}. \exists E' \in C. E \xrightarrow{a, \tilde{\lambda}, (\tilde{y}, \tilde{b})}_1 E' \land PL(\tilde{\lambda}) = l |\!\}$$
$$Yield_1(E, a, l, \tilde{b}, C) = \sum\{\!| \tilde{y} \mid \exists \tilde{\lambda}. \exists E' \in C. E \xrightarrow{a, \tilde{\lambda}, (\tilde{y}, \tilde{b})}_1 E' \land PL(\tilde{\lambda}) = l |\!\}$$
with $RY_1(E, a, l, \tilde{b}, C) = \bot$ *whenever the multisets above are empty.* ∎

Definition 7. *An equivalence relation* $\mathcal{B} \subseteq \mathcal{G}_1 \times \mathcal{G}_1$ *is a RY-Markovian bisimulation of order 1 if and only if, whenever* $(E_1, E_2) \in \mathcal{B}$, *then for all* $a \in AType$, $l \in \mathbb{Z}$, $\tilde{b} \in ARew$, *and equivalence classes* $C \in \mathcal{G}_1/\mathcal{B}$
$$RY_1(E_1, a, l, \tilde{b}, C) = RY_1(E_2, a, l, \tilde{b}, C)$$ ∎

Table 2. Axiomatization of $\sim_{MB_1^{RY}}$

$(\mathcal{A}_1)_1^{RY}$ $\qquad (E_1 + E_2) + E_3 = E_1 + (E_2 + E_3)$

$(\mathcal{A}_2)_1^{RY}$ $\qquad E_1 + E_2 = E_2 + E_1$

$(\mathcal{A}_3)_1^{RY}$ $\qquad E + \underline{0} = E$

$(\mathcal{A}_4)_1^{RY}$ $\qquad <a, \tilde{\lambda}_1, (\tilde{y}_1, \bar{b})>.E + <a, \tilde{\lambda}_2, (\tilde{y}_2, \bar{b})>.E = <a, \tilde{\lambda}_1 + \tilde{\lambda}_2, (\tilde{y}_1 + \tilde{y}_2, \bar{b})>.E$
$\qquad\qquad\qquad\qquad\qquad$ if $PL(\tilde{\lambda}_1) = PL(\tilde{\lambda}_2)$

$(\mathcal{A}_5)_1^{RY}$ $\quad <a, *_{l_1, w_1}, (*, *)>.E_1 + <a, *_{l_2, w_2}, (*, *)>.E_2 = <a, *_{l_1, w_1}, (*, *)>.E_1$
$\qquad\qquad\qquad\qquad\qquad$ if $l_1 > l_2$

$(\mathcal{A}_6)_1^{RY}$ $\qquad \underline{0}/L = \underline{0}$

$(\mathcal{A}_7)_1^{RY}$ $\qquad (<a, \tilde{\lambda}, (\tilde{y}, \bar{b})>.E)/L = <a, \tilde{\lambda}, (\tilde{y}, \bar{b})>.(E/L)$
$\qquad\qquad\qquad\qquad\qquad$ if $a \notin L$

$(\mathcal{A}_8)_1^{RY}$ $\qquad (<a, \tilde{\lambda}, (\tilde{y}, \bar{b})>.E)/L = <\tau, \tilde{\lambda}, (\tilde{y}, \bar{b})>.(E/L)$
$\qquad\qquad\qquad\qquad\qquad$ if $a \in L$

$(\mathcal{A}_9)_1^{RY}$ $\qquad (E_1 + E_2)/L = E_1/L + E_2/L$

$(\mathcal{A}_{10})_1^{RY}$ $\qquad \underline{0}[\varphi] = \underline{0}$

$(\mathcal{A}_{11})_1^{RY}$ $\qquad (<a, \tilde{\lambda}, (\tilde{y}, \bar{b})>.E)[\varphi] = <\varphi(a), \tilde{\lambda}, (\tilde{y}, \bar{b})>.(E[\varphi])$

$(\mathcal{A}_{12})_1^{RY}$ $\qquad (E_1 + E_2)[\varphi] = E_1[\varphi] + E_2[\varphi]$

$(\mathcal{A}_{13})_1^{RY}$ $\displaystyle\sum_{i \in I_0} <a_i, \tilde{\lambda}_i, (\tilde{y}_i, \bar{b}_i)>.E_i \parallel_S \sum_{i \in I_1} <a_i, \tilde{\lambda}_i, (\tilde{y}_i, \bar{b}_i)>.E_i =$

$\displaystyle\sum_{j \in I_0, a_j \notin S} <a_j, \tilde{\lambda}_j, (\tilde{y}_j, \bar{b}_j)>.(E_j \parallel_S \sum_{i \in I_1} <a_i, \tilde{\lambda}_i, (\tilde{y}_i, \bar{b}_i)>.E_i) +$

$\displaystyle\sum_{j \in I_1, a_j \notin S} <a_j, \tilde{\lambda}_j, (\tilde{y}_j, \bar{b}_j)>.(\sum_{i \in I_0} <a_i, \tilde{\lambda}_i, (\tilde{y}_i, \bar{b}_i)>.E_i \parallel_S E_j) +$

$\displaystyle\sum_{k \in K_0} \sum_{h \in P_{1, a_k}} <a_k, \tilde{\lambda}_k \cdot (w_h/W_{1, a_k}), (\tilde{y}_k \cdot (w_h/W_{1, a_k}), \bar{b}_k)>.(E_k \parallel_S E_h) +$

$\displaystyle\sum_{k \in K_1} \sum_{h \in P_{0, a_k}} <a_k, \tilde{\lambda}_k \cdot (w_h/W_{0, a_k}), (\tilde{y}_k \cdot (w_h/W_{0, a_k}), \bar{b}_k)>.(E_h \parallel_S E_k) +$

$\displaystyle\sum_{k \in P_0'} \sum_{h \in P_{1, a_k}} <a_k, *_{\max(l_k, l_h), (w_k/W_{0, a_k}) \cdot (w_h/W_{0, a_k}) \cdot N_{a_k}}, (*, *)>.(E_k \parallel_S E_h)$

where $I_0 \cap I_1 = \emptyset$, $\tilde{\lambda}_i = *_{l_i, w_i}$ for $i \in I_0 \cup I_1$. $PL(\tilde{\lambda}_k) < 0$, and for $j \in \{0, 1\}$

$\qquad L_{j,a} = \max\{l_k \mid k \in I_j \wedge a_k = a \wedge \tilde{\lambda}_k = *_{l_k, w_k}\}$

$\qquad P_{j,a} = \{k \in I_j \mid a_k = a \wedge \tilde{\lambda}_k = *_{l_k, w_k} \wedge l_k = L_{j,a}\}$

$\qquad K_j = \{k \in I_j \mid a_k \in S \wedge PL(\tilde{\lambda}_k) \geq 0 \wedge P_{1-j, a_k} \neq \emptyset\}$

$\qquad P_0' = \{k \in I_0 \mid \exists a \in S. k \in P_{0,a} \wedge P_{1,a} \neq \emptyset\}$

$\qquad W_{j,a} = \sum \{\!\!\!| \; w_k \mid k \in P_{j,a} \wedge \tilde{\lambda}_k = *_{l_k, w_k} \; |\!\!\!\}$

$\qquad N_a = \begin{cases} W_{0,a} + W_{1,a} & \text{if } L_{0,a} = L_{1,a} \\ W_{0,a} & \text{if } L_{0,a} > L_{1,a} \\ W_{1,a} & \text{if } L_{1,a} > L_{0,a} \end{cases}$

It is easy to see that the union of all the RY-Markovian bisimulations of order 1 is a RY-Markovian bisimulation of order 1. Such a union, denoted $\sim_{MB_1^{RY}}$, is called the RY-Markovian bisimulation equivalence of order 1. $\sim_{MB_1^{RY}}$ is a congruence w.r.t. all the algebraic operators as well as recursive constant definitions. In Table 2 we report from [3] the sound and complete axiomatization of nonrecursive $\text{EMPA}_{\text{gr}_1}$ terms w.r.t. $\sim_{\text{MB}_1^{\text{RY}}}$.

4 Aggregating Bonus Rewards

As witnessed by axiom $(\mathcal{A}_4)_1^{\text{RY}}$, $\sim_{\text{MB}_1^{\text{RY}}}$ aggregates rates and yield rewards without manipulating bonus rewards at all. However, if we look at the way performance measures are computed according to formulas (1) and (2), we note that whenever two actions are merged into a single one, then their bonus rewards can be aggregated as well. Unlike the rates and the yield rewards of those two actions, the bonus rewards are not just summed up as each of them needs to be preliminarily multiplied by the execution probability of the corresponding action. As an example, $<a, \lambda_1, (y_1, b_1)>.E + <a, \lambda_2, (y_2, b_2)>.E$ can be equated to $<a, \lambda_1+\lambda_2, (y_1+y_2, \frac{\lambda_1}{\lambda_1+\lambda_2}\cdot b_1 + \frac{\lambda_2}{\lambda_1+\lambda_2}\cdot b_2)>.E$; similarly $<a, \infty_{l,w_1}, (y_1, b_1)>.E + <a, \infty_{l,w_2}, (y_2, b_2)>.E$ can be equated to $<a, \infty_{l,w_1+w_2}, (y_1 + y_2, \frac{w_1}{w_1+w_2}\cdot b_1 + \frac{w_2}{w_1+w_2}\cdot b_2)>.E$. To be more precise, the probability by which each bonus reward involved in the aggregation must be multiplied is the probability of executing the corresponding action conditioned on the fact that one of the actions involved in the aggregation is executed. Considering such a conditional execution probability instead of just the execution probability is not only necessary to preserve the value of the performance measures according to formulas (1) and (2), but is also crucial to get the congruence property. We introduce below an improved reward based Markovian bisimulation congruence that aggregates bonus rewards as well.

Definition 8. *We extend the real number division to rates of the same priority level as follows:*

$$*_{l,w_1} \,/\, *_{l,w_2} \ = *_{l, w_1/w_2}$$
$$\infty_{l,w_1} \,/\, \infty_{l,w_2} = w_1/w_2$$

and we extend the real number multiplication to passive rates and unspecified rewards as follows:

$$*_{l,w} \cdot * = *$$

We define partial function aggregated rate-yield-bonus $RYB_1 : \mathcal{G}_1 \times AType \times \mathbb{Z} \times \mathcal{P}(\mathcal{G}_1) \nrightarrow ARate \times ARew \times ARew$ by:

$RYB_1(E, a, l, C) = (Rate_1(E, a, l, C), Yield_1(E, a, l, C), Bonus_1(E, a, l, C))$
where:

$$Rate_1(E, a, l, C) = \sum \{\!| \ \tilde{\lambda} \mid \exists \tilde{y}, \tilde{b}. \ \exists E' \in C. \ E \xrightarrow{a, \tilde{\lambda}, (\tilde{y}, \tilde{b})}_1 E' \land PL(\tilde{\lambda}) = l \ |\!\}$$

$$Yield_1(E, a, l, C) = \sum \{\!| \ \tilde{y} \mid \exists \tilde{\lambda}, \tilde{b}. \ \exists E' \in C. \ E \xrightarrow{a, \tilde{\lambda}, (\tilde{y}, \tilde{b})}_1 E' \land PL(\tilde{\lambda}) = l \ |\!\}$$

$$Bonus_1(E, a, l, C) = \sum \{\!| \ \frac{\tilde{\lambda}}{Rate_1(E,a,l,C)} \cdot \tilde{b} \mid \exists \tilde{y}. \ \exists E' \in C. \ E \xrightarrow{a, \tilde{\lambda}, (\tilde{y}, \tilde{b})}_1 E' \land PL(\tilde{\lambda}) = l \ |\!\}$$

with $RYB_1(E, a, l, C) = \perp$ whenever the multisets above are empty. ∎

Definition 9. *An equivalence relation* $\mathcal{B} \subseteq \mathcal{G}_1 \times \mathcal{G}_1$ *is a RYB-Markovian bisimulation of order 1 if and only if, whenever* $(E_1, E_2) \in \mathcal{B}$*, then for all* $a \in AType$*,* $l \in \mathbb{Z}$*, and equivalence classes* $C \in \mathcal{G}_1/\mathcal{B}$
$$RYB_1(E_1, a, l, C) = RYB_1(E_2, a, l, C)$$
■

It is easy to see that the union of all the RYB-Markovian bisimulations of order 1 is a RYB-Markovian bisimulation of order 1. Such a union, denoted $\sim_{\mathrm{MB}_1^{\mathrm{RYB}}}$, is called the RYB-Markovian bisimulation equivalence of order 1.

Theorem 1. $\sim_{\mathrm{MB}_1^{\mathrm{RYB}}}$ *is a congruence w.r.t. all the algebraic operators as well as recursive constant definitions.*

Proof. See [4].
■

Theorem 2. *Let* $\mathcal{A}_1^{\mathrm{RYB}}$ *be the set of axioms obtained from those in Table 2 by replacing* $(\mathcal{A}_4)_1^{\mathrm{RY}}$ *with*
$$(\mathcal{A}_4)_1^{\mathrm{RYB}} \ <a, \tilde{\lambda}_1, (\tilde{y}_1, \tilde{b}_1)>.E + <a, \tilde{\lambda}_2, (\tilde{y}_2, \tilde{b}_2)>.E =$$
$$<a, \tilde{\lambda}_1 + \tilde{\lambda}_2, (\tilde{y}_1 + \tilde{y}_2, \tfrac{\tilde{\lambda}_1}{\tilde{\lambda}_1 + \tilde{\lambda}_2} \cdot \tilde{b}_1 + \tfrac{\tilde{\lambda}_2}{\tilde{\lambda}_1 + \tilde{\lambda}_2} \cdot \tilde{b}_2)>.E$$
if $PL(\tilde{\lambda}_1) = PL(\tilde{\lambda}_2)$*. The deductive system* $Ded(\mathcal{A}_1^{\mathrm{RYB}})$ *is sound and complete for* $\sim_{\mathrm{MB}_1^{\mathrm{RYB}}}$ *over the set of nonrecursive terms of* $\mathcal{G}_1$*.*

Proof. See [4].
■

Theorem 3. *Let* $E_1, E_2 \in \mathcal{E}_1$*. If* $E_1 \sim_{\mathrm{MB}_1^{\mathrm{RYB}}} E_2$ *then the value of the reward based performance measure is the same for* E_1 *and* E_2*.*

Proof. See [4].
■

5   Mixing Yield and Bonus Rewards

Having the objective of defining a reward based Markovian bisimulation congruence that aggregates as much as possible, the question arises as to whether it is possible to consider just one type of reward instead of two. From the point of view of an equivalence, this can be rephrased in terms of being able to jointly consider yield and bonus rewards. By looking at formulas (1) and (2) and the way transition frequencies are computed, we note that in the continuous time case $<a, \lambda_1, (y_1, b_1)>.E + <a, \lambda_2, (y_2, b_2)>.E$ can be equated to $<a, \lambda_1 + \lambda_2, (y_1 + y_2 + \lambda_1 \cdot b_1 + \lambda_2 \cdot b_2, 0)>.E$, while in the discrete time case $<a, \infty_{l,w_1}, (y_1, b_1)>.E + <a, \infty_{l,w_2}, (y_2, b_2)>.E$ can be equated to $<a, \infty_{l,w_1+w_2}, (y_1 + y_2 + \tfrac{w_1}{w_1+w_2} \cdot b_1 + \tfrac{w_2}{w_1+w_2} \cdot b_2, 0)>.E$. This gives rise to a normal form where only yield rewards are actually present, with bonus rewards being zero (or unspecified in the case of passive actions). However, in the discrete time case, we observe that the factor by which each bonus reward must be multiplied is equal to the execution probability of the transition to which it is attached. Since such a probability varies depending on the context in which the term is placed, compositionality is lost. As an example, if we call E_1 and E_2 the two equivalent terms above, respectively, and we take E_3 defined by $<a, \infty_{l,w_3}, (y_3, b_3)>.E$, we have that the aggregated yield reward for $E_1 + E_3$ is $y_1 + y_2 + y_3 + \frac{w_1}{w_1 + w_2 + w_3} \cdot b_1 + \frac{w_2}{w_1 + w_2 + w_3} \cdot b_2 + \frac{w_3}{w_1 + w_2 + w_3} \cdot b_3$, while the aggregated yield reward for $E_2 + E_3$ is $y_1 + y_2 + \frac{w_1}{w_1 + w_2} \cdot b_1 + \frac{w_2}{w_1 + w_2} \cdot b_2 + y_3 + \frac{w_3}{w_1 + w_2 + w_3} \cdot 0 + \frac{w_3}{w_1 + w_2 + w_3} \cdot b_3$. We introduce below a further improved reward based Markovian bisimulation congruence that mixes yield and bonus rewards in the continuous time case.

Definition 10. *We define partial function aggregated rate-reward* $RR_1 : \mathcal{G}_1 \times AType \times \mathbb{Z} \times \mathcal{P}(\mathcal{G}_1) \rightarrow (ARate \times ARew) \cup (ARate \times ARew \times ARew)$ *by:*
$$RR_1(E, a, l, C) = \begin{cases} (Rate_1(E, a, l, C), Reward_1(E, a, C)) & \text{if } l = 0 \\ RYB_1(E, a, l, C) & \text{if } l \neq 0 \end{cases}$$
where:
$$Reward_1(E, a, C) = \sum \{\!| \tilde{y} + \tilde{\lambda} \cdot \tilde{b} \mid \exists E' \in C. E \xrightarrow{a, \tilde{\lambda}, (\tilde{y}, \tilde{b})}_1 E' \wedge PL(\tilde{\lambda}) = 0 |\!\}$$
with $RR_1(E, a, l, C) = \bot$ *whenever the multisets above are empty.* ∎

Definition 11. *An equivalence relation* $\mathcal{B} \subseteq \mathcal{G}_1 \times \mathcal{G}_1$ *is a RR-Markovian bisimulation of order 1 if and only if, whenever* $(E_1, E_2) \in \mathcal{B}$, *then for all* $a \in AType$, $l \in \mathbb{Z}$, *and equivalence classes* $C \in \mathcal{G}_1/\mathcal{B}$
$$RR_1(E_1, a, l, C) = RR_1(E_2, a, l, C)$$
∎

It is easy to see that the union of all the RR-Markovian bisimulations of order 1 is a RR-Markovian bisimulation of order 1. Such a union, denoted $\sim_{\mathrm{MB}_1^{\mathrm{RR}}}$, is called the RR-Markovian bisimulation equivalence of order 1.

Theorem 4. $\sim_{\mathrm{MB}_1^{\mathrm{RR}}}$ *is a congruence w.r.t. all the algebraic operators as well as recursive constant definitions.*

Proof. See [4]. ∎

Theorem 5. *Let* $\mathcal{A}_1^{\mathrm{RR}}$ *be the set of axioms obtained from* $\mathcal{A}_1^{\mathrm{RYB}}$ *by adding*
$$(\mathcal{A}_{4'})_1^{\mathrm{RR}} \quad <a, \lambda, (y, b)>.E = <a, \lambda, (y + \lambda \cdot b, 0)>.E$$

The deductive system $Ded(\mathcal{A}_1^{\mathrm{RR}})$ *is sound and complete for* $\sim_{\mathrm{MB}_1^{\mathrm{RR}}}$ *over the set of nonrecursive terms of* $\mathcal{G}_1$.

Proof. See [4]. ∎

Theorem 6. *Let* $E_1, E_2 \in \mathcal{E}_1$. *If* $E_1 \sim_{\mathrm{MB}_1^{\mathrm{RR}}} E_2$ *then the value of the reward based performance measure is the same for* E_1 *and* E_2.

Proof. See [4]. ∎

6   Conclusion

In this paper we have improved the performance measure sensitive Markovian bisimulation congruence of [3] in order to aggregate more states and transitions while preserving compositionality and the values of the performance measures. While the congruence of [3] aggregates yield rewards without manipulating bonus rewards at all, the congruence of Def. 11 aggregates also the bonus rewards, provided that they are multiplied by the conditional probabilities of executing the actions to which they are attached, and allows yield rewards and bonus rewards to be used interchangeably in the continuous time case, provided that they are divided/multiplied by the rates of the actions to which they are attached.

The impossibility result of this paper, i.e. the fact that it is not possible to define a performance measure sensitive Markovian bisimulation congruence that allows yield and bonus rewards to be used interchangeably in the discrete time case, emphasizes the necessity of the bonus rewards. In the literature of Markov reward processes it is well known that yield and bonus rewards can be used interchangeably in the continuous time case, and in this paper we have verified that such a property does not violate compositionality. In the continuous time case the yield rewards work well because of the race policy. In particular, the additivity assumption is sound because in every state all the transitions are viewed as being in parallel execution, hence each of them contributes with its yield reward to the accumulation of reward at the state. In the discrete time case, instead, the preselection policy applies, hence the bonus rewards are more natural to express performance measures. Besides being more convenient from the modeling viewpoint, in the discrete time case the bonus rewards are also necessary from the compositionality viewpoint, i.e. they cannot be transformed into yield rewards if we want to get a congruence. In fact, if we transform them into yield rewards, we have that the contribution of the transitions to the accumulation of reward at the state is given by their average bonus reward, i.e. the weighted sum of their bonus rewards with each of them multiplied by the execution probability of the corresponding transition. Since the above mentioned execution probabilities (unlike the rates in the continuous time case) vary depending on the environment in which the state is placed, compositionality is lost.

The performance measure sensitive Markovian bisimulation congruence of Def. 11 aggregates more than that of [3]. The reason is that the new congruence can merge also those transitions that the old congruence cannot merge only because of their different bonus rewards. A quantification of the achieved improvement is left for future research.

We conclude with a practice related observation. Describing the rewards directly within the Markovian process algebra specifications of the systems has the advantage of allowing the specifications to be compositionally manipulated while preserving the values of the performance measures. This advantage on the analysis side is unfortunately diminished by a drawback on the modeling side: the system specifications are obfuscated with performance measure related details. However, the Markovian process algebra specification of a system and the specification of its reward based performance measures of interest can be easily decoupled by separately describing the rewards to be attached to the actions occurring in the system specification. A syntactical preprocessing step, like the one performed by the EMPA$_{gr_n}$ based software tool TwoTowers [2], then permits to automatically insert the rewards into the system specification. This avoids burdening the system specifications with rewards at modeling time, eases the specification of additional performance measures for the same system specification, and allows every performance measure specification to be reused for different system specifications.

References

1. C. Baier, J.-P. Katoen, H. Hermanns, *"Approximate Symbolic Model Checking of Continuous Time Markov Chains"*, in Proc. of CONCUR '99, LNCS 1664:146-162
2. M. Bernardo, *"Theory and Application of Extended Markovian Process Algebra"*, Ph.D. Thesis, University of Bologna (Italy), 1999
   (http://www.di.unito.it/~bernardo/)
3. M. Bernardo, M. Bravetti, *"Performance Measure Sensitive Congruences for Markovian Process Algebras"*, to appear in Theoretical Computer Science, 2001
4. M. Bernardo, M. Bravetti, *"Formal Specification of Performance Measures for Process Algebra Models of Concurrent Systems"*, Tech. Rep. UBLCS-1998-08, University of Bologna (Italy), 1998 (revised 2001)
5. M. Bravetti, M. Bernardo, *"Compositional Asymmetric Cooperations for Process Algebras with Probabilities, Priorities, and Time"*, Tech. Rep. UBLCS-2000-01, University of Bologna (Italy), 2000 (extended abstract in MTCS '00, Electronic Notes in Theoretical Computer Science 39(3))
6. G. Clark, *"Formalising the Specification of Rewards with PEPA"*, in Proc. of PAPM '96, CLUT, pp. 139-160
7. G. Clark, S. Gilmore, J. Hillston, *"Specifying Performance Measures for PEPA"*, in Proc. of ARTS '99, LNCS 1601:211-227
8. R.J. van Glabbeek, S.A. Smolka, B. Steffen, *"Reactive, Generative and Stratified Models of Probabilistic Processes"*, in Information and Computation 121:59-80, 1995
9. B.R. Haverkort, K.S. Trivedi, *"Specification Techniques for Markov Reward Models"*, in Discrete Event Dynamic Systems: Theory and Applications 3:219-247, 1993
10. R.A. Howard, *"Dynamic Probabilistic Systems"*, John Wiley & Sons, 1971
11. V.F. Nicola, *"Lumping in Markov Reward Processes"*, Tech. Rep. RC-14719, IBM T.J. Watson Research Center, Yorktown Heights (NY), 1990
12. W.H. Sanders, J.F. Meyer, *"A Unified Approach for Specifying Measures of Performance, Dependability, and Performability"*, in Dependable Computing and Fault Tolerant Systems 4:215-237, 1991
13. J.E. Voeten, *"Temporal Rewards for Performance Evaluation"*, in Proc. of PAPM '00, Carleton Scientific, pp. 511-522

Using Max-Plus Algebra for the Evaluation of Stochastic Process Algebra Prefixes

Lucia Cloth, Henrik Bohnenkamp, and Boudewijn Haverkort

Department of Computer Science
Laboratory for Performance Evaluation and Distributed Systems
RWTH Aachen, D-52056 Aachen
{lucia, henrik, haverkort}@cs.rwth-aachen.de
Phone: +49 241 44021432

Abstract. In this paper, the concept of complete finite prefixes for process algebra expressions is extended to stochastic models. Events are supposed to happen after a delay that is determined by random variables assigned to the preceding conditions. Max-plus algebra expressions are shown to provide an elegant notation for stochastic prefixes not containing any decisions. Furthermore, they allow for the computation of performance measures. The derivation of the so called k-th occurrence times is shown in detail.

1 Introduction

Stochastic process algebras (SPA) have become accepted languages for the description of functional and quantitative aspects of distributed systems. Their compositionality allows for easy-to-read specifications and for the reuse of modules. The evaluation and verification of SPA models is frequently based on interleaving semantics, which are prone to state-space explosion. Action delays are often restricted to exponential distributions, so that results can be obtained by Markovian analysis. Most interleaving semantics do not allow for general distributions.

In this paper, we avoid some of the above restrictions by using a true-concurrency semantics: a finite stochastic event structure prefix. Because of its non-interleaving nature, it is generally smaller than an interleaving transition system, and has an explicit notion for the concurrent execution of actions. Additionally, it allows for general distributions for the description of action delays. Our semantics for a simple stochastic process algebra is based on the complete finite prefix for (non-stochastic) process algebra introduced in [9], which in turn was inspired by a McMillan prefix for Petri nets [10] (improved by [5]). Expressions in max-plus algebra [2] are shown to be a natural description for timing properties of systems. We show that our approach is appropriate for the derivation of performance measures. Similar work can be found in [13].

The rest of this paper is organised as follows. In Section 2 we briefly review the derivation of a complete finite prefix for simple stochastic process algebra

L. de Alfaro and S. Gilmore (Eds.): PAPM-PROBMIV 2001, LNCS 2165, pp. 152–167, 2001.

expressions. Section 3 gives an overview of max-plus algebra, especially of max-plus matrix operations and their application to graphs and systems of linear equations. We use max-plus methods in Section 4 for the description and evaluation of prefixes. Section 5 concludes with a discussion of the results, drawbacks and further ideas. The material presented in this paper is based on [4].

2   Complete Finite Prefix for Stochastic Process Algebra

This section gives a brief outline on computing prefixes. Further details can be found in [4,9].

2.1   Stochastic Process Algebra

We use a simple SPA throughout this paper. Its syntax is defined as follows.

Definition 1. *Let a be an action, F a distribution function with $F(x) = 0$ for $x < 0$, and A a set of actions. The* **stochastic process algebra expressions** *are defined by the following rules:*

$$P \quad ::= \quad \text{stop} \quad | \quad \langle a, F \rangle.P \quad | \quad P + P \quad | \quad P\|_A P \quad | \quad Id \qquad \diamond$$

stop is the process that does nothing. $\langle a, F \rangle.P$ is the process that executes action a after a delay that is distributed according to F. We can think of a clock that is set to a randomly chosen delay (according to F) and that counts downwards to 0, where it eventually expires. After the clock has expired, action a can be executed. The execution of an action does not consume time. The process $P_1 + P_2$ may behave either as P_1 or as P_2. Which behaviour is chosen depends on the clocks that are running for P_1 and P_2: the fastest wins. If the winner can not be uniquely determined, the choice is made non-deterministically. The process $P_1\|_A P_2$ describes the independent parallel execution of P_1 and P_2. Only if one of them wants to execute an action that is contained in the synchronisation set A, it has to wait until the other process becomes ready to execute this action as well. Both execute the action synchronously. An identifier (Id) is a place-holder for an expression P that is defined by an equation $Id = P$. We can instantiate a process several times, i.e., for recursive definitions.

Example 1. As running example, we consider a simple buffer with a writing and a reading agent. Data of random length is written into the buffer (in) and later read from the buffer (out). In order to reflect the random length, both operations require an exponential delay, but reading is faster than writing.

The writer thinks for a uniformly distributed amount of time (between 1 and 5 time units) and writes then something into the buffer:

$$\text{Writer} = \langle \text{thinkW}, \text{uniform}([1,5]) \rangle.\langle \text{in}, \exp(5) \rangle.\text{Writer}$$

The reader waits exactly 2 time units before it reads from the buffer:

$$\text{Reader} = \langle \text{thinkR}, \det(2) \rangle.\langle \text{out}, \exp(8) \rangle.\text{Reader}$$

with
$$\text{Buffer} = \langle \text{in}, \det(0) \rangle . \langle \text{out}, \det(0) \rangle . \text{Buffer}.$$

The entire system is described by

$$\text{System} = \text{Writer} \|_{\{\text{in}\}} \text{Buffer} \|_{\{\text{out}\}} \text{Reader}.$$

Note that the buffer itself does not delay the Writer nor the Reader upon in and out actions, respectively, since its delay is 0 time units with probability 1. □

2.2   Condition Event Structures

The base structure for prefixes are *condition event structures*. Events describe the possible occurrences of actions. Each possible occurrence of an action is denoted by an unique event. We will here identify events by their corresponding action name, possibly adding a subscript in order to form a unique name.

Definition 2. *A* **stochastic condition event structure** *is a tuple* $(D, E, \sharp, \prec, l_{\mathcal{F}})$ *where*

- *D is a set of* **conditions**,
- *E is a set of* **events**,
- *$\sharp \subset D \times D$ is the* **choice relation** *(symmetric and irreflexive)*
- *$\prec \subset (D \times E) \cup (E \times D)$ is the* **flow relation**,
- *$l_{\mathcal{F}} : D \longrightarrow DF$ is a mapping from conditions to distribution functions.*   ◇

Note that neither D nor E has to be finite. For technical reasons, we assume E to contain a bottom event $\perp$ that denotes the start of the modelled system behaviour.

Stochastic condition event structures have a simple graphical representation. Conditions are drawn as circles, labelled with their names and distributions. Events simply appear with their names. The flow relation is represented by arrows, the choice relation by lines with a $\sharp$ on them.

An event structure models all possible behaviours of a system. With the start of the system, we also set imaginary clocks assigned to the conditions directly following the bottom event $\perp$, i.e., for all $d \in D : \perp \prec d$. If for any event the clocks of all directly preceding conditions have expired, this event occurs immediately, starting the clocks of the succeeding conditions, and so on.

The occurrence of an event e inhibits forever the occurrence of all events being in conflict with e.

Downwards closed sets of pairwise non-conflicting events are called *configurations*. The *local configuration* $[e]$ of an event e is defined by

$$[e] := \{e' \in E | e' \prec^* e\}.$$

$\emptyset$ is the local configuration of $\perp$. Corresponding to configurations are *cuts*: a cut is the set of conditions "following" a configuration. Each configuration corresponds to a cut and vice versa. After the occurrence of an event, usually some time has to pass before the next occurrence of an event. Consequently, cuts represent *states* in which the system may spend time.

2.3   Unfolding

The algorithm in [9] for the construction of a condition event structure for a process algebra expression is derived from the unfolding algorithm for Petri nets [10]. So we refer to the resulting event structure as *unfolding* as well. As shown in [4], the unfolding algorithm in [9] can directly be used to derive unfoldings for SPA expression, yielding stochastic event structures.

The unfolding algorithm, which we do not describe in detail here, "unfolds" the SPA expression, generating conditions and events step by step. The conditions are additionally labelled with so-called *components* (via a mapping l_C). Components are basically prefixed stochastic process algebra expressions, equipped with a notion for the synchronisation context in which they occur.

Definition 3. *Let P be a stochastic process algebra expression, a an action, and A a set of actions and F a distribution function. A **component** C is defined by*

$$C \quad ::= \quad \text{stop} \quad | \quad \langle a, F \rangle.P \quad | \quad C\|_A \quad | \quad \|_A C.$$

$\diamond$

Stochastic process algebra expression are decomposed into components (roughly by splitting parallel and choice expressions).

Components are assigned to conditions. The choice relation between components, derived during the decomposition of SPA expressions, determines the choice relation between the respective conditions.

For sets of components a structured operational semantics (SOS) can be defined [9]. Starting from a set of initial components, all possible transitions (labelled with actions) are derived according to this SOS. For each transition, a new event and its successor conditions (with appropriately assigned components) is introduced in the unfolding. Hence, each occurrence of an action is transformed into a unique event. The whole unfolding is created by successive application of this derivation of transitions from "unused" components in the unfolding.

If an action is executed by a single component, the corresponding event will have exactly one preceding condition; if it is the result of a synchronisation, it will have several predecessors. Apart from the explicit choice, described by the choice relation on components, there can also be choice as a result of synchronisation on common labels that can be recognised by the existence of conditions with more than one succeeding event. In general (for recursive SPA expressions) there is an infinite number of possible occurrences of actions, and so the resulting stochastic condition event structure will be an infinite structure.

States. A cut in the unfolding corresponds to a set of conditions, and so it does (via the mapping l_C) to a set of components. These sets will be such that they correspond to a valid stochastic process algebra expression that is reachable (via the classical interleaving semantics) from the original SPA expression. The expressions can be assembled by undoing the decomposition. They are called the *states* of the cut (resp. of the corresponding configuration). In [9] it is shown that *exactly* the reachable states are represented in the unfolding.

2.4   Prefix

For finite state SPA processes, the possible behaviour is already represented in a finite prefix of the unfolding. In this section we comment on the construction of the prefix.

Events are made comparable by a so-called *adequate order* $\sqsubseteq$ such that (roughly) $e' \sqsubseteq e$ if e' is encountered earlier in the unfolding process. There is a local configuration $[e]$ for all events e in the unfolding and consequently we have local states, written $St([e])$. We use the idea of states assigned to events for the construction of a finite prefix: Whenever we find an event e, such that there exists an event e' with $e' \sqsubseteq e$ and $St([e]) = St([e'])$ we call it a *cutoff* event. We do not consider the unfolding beyond the cuts of cutoff events. The resulting prefix is finite and complete (although not always the smallest possible).

The events in the prefix can be seen as representatives of whole classes of events. They represent the finite number of possible actions, the occurrences of which determine the event classes. These classes may have an infinite number of elements.

Example 2. In Figure 1 the unfolding and the complete finite prefix for the SPA expression System of Example 1 is depicted. In order to distinguish different occurrences of an action, events have action names with unique subscripts. We name conditions with capital letters.   □

3   Max-Plus Algebra

3.1   Introduction

The so-called max-plus algebra has been developed for the description and evaluation of discrete-event systems (DES). A complete survey can be found in [2]. Stochastic DES are treated in [12]. The max-plus algebra is an algebraic ring comprising the set of real numbers (extended by $-\infty$) as carrier set and $\oplus$ and $\otimes$ as operations. $\oplus$ is interpreted as max and $\otimes$ as $+$ (we also write $\oslash$ for $-$). In our models we express time by random variables that are mappings from some probability space into the set of real numbers. By $\mathbf{0}$ we denote the random variable which is $-\infty$ with probability 1; it is the zero-element for the maximum operation $\oplus$. With $\mathbb{1}$ we denote the random variable which is 0 with probability 1; it is the unit-element for $\otimes$.

3.2   Matrices and Weighted Directed Graphs

The operations $\oplus$ and $\otimes$ can be extended to vectors and matrices whose elements are drawn from the max-plus algebra.

Definition 4. *A **directed graph** (digraph) $G = (V, E)$ consists of a set $V = \{1, 2, \ldots, n\}$ of nodes (vertices) and a set $E \subseteq V \times V$ of arcs (edges).*

    *G is a max-plus **weighted** digraph, if it comes with a mapping w that assigns to each arc a max-plus random variable.*   ◇

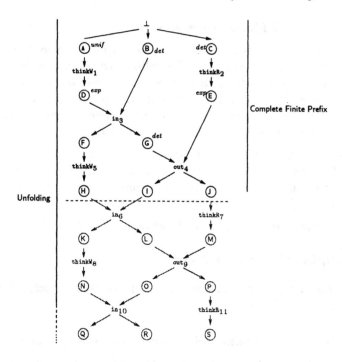

Complete Finite Prefix

Unfolding

Fig. 1. Unfolding and Complete Finite Prefix for **Writer** $\|_{\text{in}}$ **Buffer** $\|_{\text{out}}$ **Reader**

An alternative representation for a graph is an *adjacency matrix* A, where

$$A_{ij} := \begin{cases} w((i,j)), & \text{if } (i,j) \in E, \\ 0, & \text{otherwise.} \end{cases}$$

Of course it is also possible to construct a graph to a given $n \times n$ max-plus matrix. Squaring the adjacency matrix A results in A^2, which represents the graph with an arc between two nodes n_1 and n_2 into the resulting graph, if there is a path of length two in the graph described by A from node n_1 to node n_2. Its weight is given by the sum of the original weights. In the *star* of a matrix, paths of arbitrary length are subsumed.

Definition 5. *For A an adjacency matrix, A^* is defined by the max-plus sum, that is the maximum, of all its powers:*

$$A^* := \bigoplus_{k=0}^{\infty} A^k.$$

◇

In an acyclic graph, there is a maximum length for all paths, and so it is possible to compute the star with a finite number of operations:

$$A^* = \bigoplus_{k=0}^{n} A^k, \quad \text{for some } n < \infty.$$

3.3   Systems of Linear Equations

We can formulate systems of linear equations in max-plus algebra, as we can do in the field of real numbers. The well-known techniques for the solution of systems of linear equations depend on the inverse of $\oplus$. Unfortunately, max has no inverse. Consequently, we are not able to solve general systems of linear equations in the max-plus algebra. However, there is a special case that is well suited for our purposes [2], as we will see in Section 4.

Theorem 1. *Consider a system of linear equations as follows: $T = T \otimes A \oplus S$, where T is the solution vector we are looking for, A is a matrix and S a given vector of suitable dimensions. If A is strictly upper triangular, the solution is given by $T = S \otimes A^*$.*

Proof. By insertion of $S \otimes A^*$ in the equation $T = T \otimes A \oplus S$.   ◇

4   Application to Prefixes

We use the max-plus methods introduced in the previous section for the notation and evaluation of prefixes.

4.1   Occurrence Times and Linear Max-Plus Expressions

The flow relation of the event structure prefixes describes the causal dependencies between events. Conditions are equipped with random variables that describe the delay between events. Combining dependencies and delays, we can determine a random variable $O(e)$ describing the time the event e is bound to occur.

We can derive these occurrence times recursively, starting with the bottom event $\perp$ which we assume to happen at time $O(\perp) = 0 = \mathbb{1}$. Then, we look at immediate successor events of $\perp$: their occurrence times are determined by the maximum delay assigned to the conditions between $\perp$ and themselves (plus the occurrence time of $\perp$, which is zero). We introduce for each condition d a random variable X_d distributed according to $l_{\mathcal{F}}(d)$. Then we can express occurrence times for all events: we add the random variable of the appropriate conditions to the occurrence times of the predecessors and then take the maximum.

Definition 6. *The occurrence time $O(e)$ of an event e is defined recursively as $O(\perp) := \mathbb{1}$ and $O(e) := \bigoplus_{e' \prec d \prec e} O(e') \otimes X_d$ for $e \neq \perp$, respectively.*   ◇

The resulting expressions are linear in the max-plus algebra.

Example 3. For the unfolding in Figure 1 we have the following occurrence times:

$$O(\mathtt{thinkW_1}) = O(\perp) \otimes X_A$$
$$O(\mathtt{thinkR_2}) = O(\perp) \otimes X_C \quad \text{deterministic}$$
$$O(\mathtt{in_3}) = (O(\mathtt{thinkW_1}) \otimes X_D) \oplus (O(\perp) \otimes X_B)$$
$$O(\mathtt{thinkW_5}) = O(\mathtt{in_3}) \otimes X_F$$
$$O(\mathtt{out_4}) = (O(\mathtt{in_3}) \otimes X_G) \oplus (O(\mathtt{thinkR_2}) \otimes X_E)$$
$$O(\mathtt{thinkR_7}) = O(\mathtt{out_4}) \otimes X_J$$

$O(\texttt{thinkW}_1)$ is a uniform random variable with expectation 3; $O(\texttt{in}_3)$ is the sum of a uniform and an exponential random variable, its expectation is 3.2. $O(\texttt{thinkR}_7)$ is the sum of a deterministic random variable and the maximum of sums of random variables. Its expectation is 5.6857. □

4.2 Representation of Prefixes

In their graphical form, prefixes have a close resemblance to max-plus weighted graphs. We now show how to construct a graph (and the corresponding matrix) for a stochastic event structure prefix.

We interpret events as the nodes of a graph. The flow relation seen at event level represents the arcs, i.e., if there is a condition d such that $e_1 \prec d \prec e_2$, then there is an arc from e_1 to e_2, denoted (e_1, e_2). The weight of this arc is given by the delay assigned to d.

Definition 7. *For a finite stochastic condition event structure $(D, F, \sharp, \prec, l_{\mathcal{F}})$ the corresponding weighted digraph G is defined as follows:*

- $V = F$,
- $(e_1, e_2) \in E : \Longleftrightarrow \exists d \in D : e_1 \prec d \prec e_2$
  and $w((e_1, e_2)) := X_d$ *with* $X \sim l_{\mathcal{F}}(d)$. ◇

Note that we are loosing information: the conflict between two events is no longer expressed!

Example 4. The graph (left) and matrix (right) of Figure 2 represent the prefix of Figure 1. □

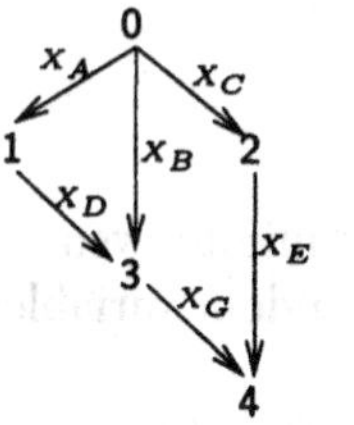

$$\begin{pmatrix} 0 & X_A & X_C & X_B & 0 \\ 0 & 0 & 0 & X_D & 0 \\ 0 & 0 & 0 & 0 & X_E \\ 0 & 0 & 0 & 0 & X_G \\ 0 & 0 & 0 & 0 & 0 \\ 0 & 0 & 0 & 0 & 0 \end{pmatrix}$$

Fig. 2. Graph and Matrix for the Prefix of Figure 1

4.3 The Repeating Part

In the preceding section we have shown how max-plus matrices can be used for the representation of prefixes, although they do not reflect the conflict situations between events. Therefore, we have to restrict ourselves to *decision-free* prefixes, i.e., prefixes without conflicts between events; all events must eventually occur. For the corresponding SPA expressions this means that they are not allowed to contain the choice operator and that they have to avoid synchronisations on common labels.

Finite prefixes are constructed by finding cutoff events in the unfolding of a stochastic process algebra expression. To any such cutoff event e in an unfolding there is another event e' that is smaller w.r.t. the chosen adequate order ($e' \sqsubset e$) and has the same local state ($St(e') = St(e)$). As a consequence, for an event $e*'$ succeeding the local configuration of e', there is an $e*$ succeeding the local configuration of e and $St(e*) = St(e*')$. Thus, by repetition of this argument we can state that the events of the prefix between e' and e, i.e., the set $[e] \setminus [e']$, represents event classes with an infinite number of members. Since in decision-free systems all events must eventually happen, the occurrence of representatives of these classes are observed infinitely often in the evolution of the system.

Definition 8. *Let Cutoff be the set of all cutoff events in a given decision-free prefix $(D, E, \emptyset, \prec, l_{\mathcal{F}}, l_C)$ and Cutoff' the set of those preceding events that have the same local state. Then the set of* **repeating events** *of the prefix is defined as*

$$E^r := \bigcup_{e \in Cutoff} [e] \setminus \bigcup_{e' \in Cutoff'} [e'].$$

The set of **repeating conditions** *is given by $D^r := \{d \in D \mid \exists e \in E^r : d \prec e \vee e \prec d\}$. The* **repeating part** *of the prefix is given by $(D^r, E^r, \emptyset, \prec \lceil (D^r \times E^r \cup E^r \times D^r))$.*   ◇

Example 5. Figure 3 (a) shows the repeating part of the prefix of Figure 1.   □

4.4   The k-th Occurrence Time

When does the k-th representative of a certain infinite event class occur? As all time intervals in our system are described by random variables, the instant can also be expressed as a random variable.

To make the formal treatment easier, we identify the events (or more precise, the event classes) E^r of the repeating part with natural numbers starting with 1, in such a way that $<$ on the natural numbers respects the partial order on events. With respect to the graphical representation, we do a *topological sorting* of events; for the running example we actually have already done this with the subscripts added to action names.

Definition 9. *For event class i, $T_i(k)$ is the instant of time when the k-th representative of this class occurs. $T(k)$ is then a max-plus vector containing the k-th occurrence times for all event classes.*   ◇

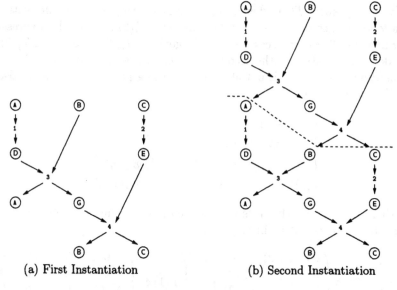

(a) First Instantiation   (b) Second Instantiation

Fig. 3. Repeating Part of the Prefix

$T(1)$ is easily determined. We simply take the original prefix and compute the occurrence times for all events in the repeating part. Put in the order given by the topological sorting, we obtain $T(1)$.

Theorem 2. *Let e_i be the minimal element of event class i according to $\prec^*$. The first occurrence time of class i is then given by the occurrence time of e_i:*

$$T_i(1) = O(e_i).$$

Proof. If $e_i \prec^* e$, then e_i must occur before e. Because of being minimal, e_i is the first representative of class i to happen. $\diamond$

For our example

$$T(1) = (X_A, X_C, X_A \otimes X_D, X_A \otimes X_D \oplus X_C \otimes X_E).$$

But how do we derive $T(k)$ for $k > 1$? We can append a new instantiation of the repeating part to the prefix by identifying (final) conditions of the prefix with equally labelled (initial) conditions of the repeating part (cf. Figure 3 (b)). Doing this infinitely often, we obtain the whole infinite unfolding.

The events of this new instantiation constitute the second occurrences of the infinite event classes. One can easily calculate their occurrence times. But there is a formal problem: we must distinguish between the random variables of conditions with identical labels from the first and the second instantiation. We do this by introducing explicitly the parameter k: $X_d(k)$ denotes the random variable associated to d in the k-th instantiation of the repeating part.

We already know $T(1)$ when computing $T(2)$. Similar to the derivation of ordinary occurrence times, we are able to describe $T_i(2)$ by max-plus expressions concerning the direct predecessors of i. In general, $T_i(2)$ depends on $T_j(2)$ for its predecessor events j in the second instance of the repeating part. If any of the conditions preceding i is initial in the repeating part, $T_i(2)$ depends also on $T(1)$. For our example

$$T_1(2) = T_3(1) \otimes X_A(2),$$
$$T_2(2) = T_4(1) \otimes X_C(2),$$
$$T_3(2) = T_1(2) \otimes X_D(2) \oplus T_4(1) \otimes X_B(2),$$
$$T_4(2) = T_3(2) \otimes X_G(2) \oplus T_2(2) \otimes X_E(2).$$

A sharp look reveals that this is a system of linear equations in the max-plus algebra. It can be rewritten in matrix form:

$$T(2) = \left(T(2) \otimes \begin{pmatrix} 0 & 0 & X_D(2) & 0 \\ 0 & 0 & 0 & X_E(2) \\ 0 & 0 & 0 & X_G(2) \\ 0 & 0 & 0 & 0 \end{pmatrix} \right) \oplus \left(T(1) \otimes \begin{pmatrix} 0 & 0 & 0 & 0 \\ 0 & 0 & 0 & 0 \\ X_A(2) & 0 & 0 & 0 \\ 0 & X_C(2) & X_B(2) & 0 \end{pmatrix} \right)$$

$T(1)$ is already known, so the second part of the sum forms a vector, hence the equation is in the general form described in Theorem 1.

The results for $T(2)$ can be generalised. The vector $T(k+1)$ depends on itself for "internal" event classes, and on $T(k)$ for "initial" event classes.

Definition 10. *The **internal dependencies within the $(k+1)$-th instantiation** of the repeating part are expressed by*

$$A_{ij}(k+1) := \begin{cases} X_d(k+1), & \text{if } i \prec d \prec j, \\ 0, & \text{otherwise}, \end{cases}$$

*The **dependencies between the k-th and the $(k+1)$-th instantiation** of the repeating part are given by*

$$B_{ij}(k+1) := \begin{cases} X_d(k+1), & \text{if } d \text{ initial}, d \prec j, d' \text{ final}, i \prec d', l_C(d) = l_C(d'), \\ 0, & \text{otherwise}, \end{cases}$$

With these definitions we get

Theorem 3.

$$T_i(k+1) = \left(\bigoplus_{j=1}^{n} T_j(k+1) \otimes A_{ij}(k+1) \right) \oplus \left(\bigoplus_{j=1}^{n} T_j(k) \otimes B_{ij}(k+1) \right)$$

In matrix form:

$$T(k+1) = \left(T(k+1) \otimes A(k+1) \right) \oplus \left(T(k) \otimes B(k+1) \right)$$

◇

Due to the acyclic topology of the repeating part and the adequate labelling of event classes, the $A(k)$ are strictly upper triangular matrices. Hence, according to Theorem 1, we state

Theorem 4.

$$T(k+1) = T(k) \otimes B(k+1) \otimes A(k+1)^*, \quad \text{for } k > 1.$$

Proof. Application of Theorem 1.   ◇

The calculation of $T(k+1)$ now simply depends on $T(k)$. Starting with $T(1)$, which is given by the occurrence times, all $T(k)$ can be determined recursively, step by step. For short, set $C(k) := B(k) \otimes A(k)^*$, such that the equation becomes

$$T(k+1) := T(k) \otimes C(k+1).$$

$B(k)$ contains the dependencies of all event classes to the previous occurrence of final events. $A(k)^*$ cumulates the internal delays such that they are expressed directly from one event to another without intermediate event. $C(k) = B(k) \otimes A(k)^*$ expresses the delay between the terminal events of instantiation number k and the $(k+1)$-th occurrence of all event classes. Thus, only the rows corresponding to terminal event classes contain entries different from 0 at all.

Example 6. For the running example, taking into account that $X_B(k)$ and $X_G(k)$ are deterministically 0, we find:

$$C(k) = \begin{pmatrix} 0 & 0 & 0 & 0 \\ 0 & 0 & 0 & 0 \\ X_A(k) & 0 & X_A(k) \otimes X_D(k) & X_A(k) \otimes X_D(k) \\ 0 & X_C(k) & 0 & X_C(k) \otimes X_E(k) \end{pmatrix}$$

□

4.5   Cycle Time

Based on the k-th occurrence times of the event classes we can derive recurrence times for the different classes. As recurrence time we designate the time between the occurrence of two subsequent representatives of a single event class. It is again a random variable, describing the distribution of time. The recurrence times of all event classes are comprised in a vector $RT(k)$:

Definition 11. $RT_i(k)$ denotes the **recurrence time** between the k-th and the $(k+1)$th occurrence of an representative of event class i:

$$RT(k) := T(k+1) \oslash T(k)$$

◇

The distributions of the recurrence times can change over time. We are interested in the long term behaviour of a system, so we take the limit of $RT(k)$ as a measure for the general recurrence times (given existence):

$$RT := \lim_{k \to \infty} RT(k).$$

The expected limiting recurrence time must be the same for all event classes. If one class were "faster" than the others, it would have to wait for its predecessors, so it would no longer be faster. Clearly this is only true for those prefixes with connected repeating part.

We take this identical limiting recurrence times as cycle time for the whole system. It denotes the time it takes to execute one instantiation of the repeating part of the system.

Definition 12. *Let CT be the expected limiting recurrence time for any event class:*

$$CT := E[RT_i], \quad \text{for arbitrarily chosen } 1 \leq i \leq n.$$

CT is called the **mean cycle time** *of the system.*   ◇

Unfortunately, the calculation of the mean cycle time comes with all those problems known from the solution of task graph models [1].

Example 7. For our example, the mean cycle time for all event classes is

$$CT = E[(X_A \otimes X_D) \oplus (X_C \otimes X_E)] = 5.6857.$$

□

4.6   Condition Holding Times

So far we presented two simple performance measures that can be derived from a complete finite prefix: the k-th occurrence times of events and the mean cycle time of the repeating part. In this section we show how to use the k-th occurrence time of events for the calculation of a performance measure concerning conditions.

Conditions are assigned stochastic delays. If the system enters a condition, a clock is started according to the specified distribution function. When this clock reaches zero, the succeeding event is locally activated. But it does not need to be the case that the event is also globally activated. Sometimes the system must remain in a condition, even when the clock already has reached zero. The amount of time the system actually spends in a condition, i.e., the delay plus the possible waiting time, is called its holding time.

Definition 13. *Let α be a non-terminal condition of the repeating part. Then*

$$H_\alpha(k) = \begin{cases} T_{j1}(k) \oslash T_{j2}(k), & \text{if } j_2 \prec \alpha \prec j_1, \\ T_{j1}(k) \oslash T_{j2}(k-1), & \text{if } \alpha \text{ initial}, \alpha \prec j_1, j_2 \prec \alpha', l_C(\alpha') = l_C(\alpha), \end{cases}$$

denotes the **holding time** *of condition α in the k-th instantiation of the repeating part.*   ◇

In a decision-free unfolding, each condition has exactly one predecessor and one successor event. The holding time in the k-th instantiation of a condition i is then easily determined: it is the difference of the k-th occurrence time of its successor and the k-th occurrence of its predecessor. If the condition is initial, the $(k-1)$-th occurrence of the predecessor event has to be considered.

The long-run holding time of a condition α is determined by the expectation of the limit of $H_\alpha(k)$, given existence:

$$H_\alpha := E\left[\lim_{k\to\infty} H_\alpha(k)\right].$$

Example 8. Since $H_A(k) = 3$ for all k, $H_A = 3$. The sum of the mean holding times for A and D has to be equal to the mean cycle time: $H_D = CT - H_A = 5.6857 - 3 = 2.6857$. Compare this to $E[X_D] = 0.2$. The mean holding time for C is $H_C = 2$. Consequently, $H_E = 3.6857$ (compared with $E[X_E] = 0.125$).

The writer is expected to think for three time units, then he is expected to wait for the buffer. This takes (on average) 2.4857 time units. Writing into the buffer takes 0.2 time units. For the reader process it is quite similar   □

4.7 Stationary Probabilities

Prefixes are a semantics for stochastic process algebra expressions. Those expressions are composed of different agents, which interact with each other. The agents are further subdivided into components, which are then used as condition labels in the prefix. We can rebuild the agents from these labels, even if we only have the prefix. An agent is then represented by a set of conditions forming a chain w.r.t. the flow relation $\prec$.

Example 9. In Fig. 4 the three agents of the repeating part of our example are depicted.   □

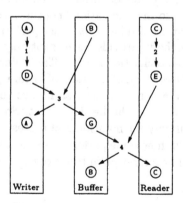

Fig. 4. The agents (Writer, Buffer and Reader) of the repeating part

For the repeating part of an agent we know the holding times of the corresponding conditions. Furthermore, we know the time the system needs for one cycle. This is exactly the same time the agent needs for one cycle. Consequently, the sum of all holding times of one agent must be the cycle time. But this means that we are able to determine the part of time the agent spends in a particular condition. In other words, we can determine the stationary probability of a condition. In contrast to stationary probabilities of Markov chains, these probabilities are local to the agent and *do not* concern the global system state.

Definition 14. *Let CT be the expected cycle time of a system and H_α the expected holding time of condition α. Then the* **local stationary probability** *of condition α is given by the fraction*

$$\pi_\alpha := \frac{H_\alpha}{CT}.$$

$\diamond$

Example 10. For the writer and the reader of the example, we obtain the following local stationary probabilities:

- The writer is thinking with probability $\pi_A = \frac{H_A}{CT} = \frac{3}{5.6857} = 0.5276$. He is waiting for or writing in the buffer with probability $\pi_D = 0.4724$.
- The reader is thinking with probability $\pi_C = 0.3518$. He is waiting for or reading from the buffer with probability $\pi_E = 0.6482$.

5   Conclusion

In this paper, we considered complete finite stochastic event structure prefixes as true-concurrency semantics for simple SPA expressions. After a short introduction in max-plus algebra, a method for the computation of occurrence times has been presented. Using the idea of cutoff events, we found the repeating event classes of a prefix. Similar to the calculation of occurrence times, k-th occurrence times can be derived that are the basis for other performance measures.

Max-plus algebra methods provide an elegant way to describe the timing behaviour of decision-free stochastic prefixes. The timing of repeating behaviour can be determined via the solution of a system of linear equations. Unfortunately, this solution is generally a complex task, as it involves sums and maxima of random variables, that even might be dependent. So, complexity may make max-plus matrices useful only for notational purposes.

Finally, we have shown how some performance measures of SPA models with *general distributions* can be expressed in terms of the stochastic prefix and max-plus algebra.

References

1. F. Baccelli, A. Jean-Marie, and Zhen Liu. A survey on solution methods for task graph models. In N. Gotz, U. Herzog, and M. Rettelbach, editors, *Proc. of the QMIPS-Workshop on Formalism, Principles and State-of-the-art. Arbeitsberichte des IMMD 26(14)*. Universität Erlangen, 1993.

2. F. L. Baccelli, G. Cohen, G. J. Olsder, and J.-P. Quadrat. *Synchronization and Linearity. An Algebra for Discrete Event Systems.* John Wiley & Sons, 1992.

3. E. Brinksma, J.-P. Katoen, R. Langerak, and D. Latella. A stochastic causality-based process algebra. *The Computer Journal*, 38(7):552–565, 1995.

4. L. Cloth. Complete finite prefix for stochastic process algebra. Master's thesis, RWTH Aachen, Department of Computer Science, 2000. available at `ftp://ftp-lvs.informatik.rwth-aachen.de/pub/da/2000/cloth.ps.gz`.

5. J. Esparza, S. Römer, and W. Vogler. An improvement of McMillan's unfolding algorithm. In T. Margaria and B. Steffen, editors, *Proc. of TACAS'96*, LNCS 1055, pages 87–106. Springer, 1996.

6. P. Glynn. A GSMP formalism for discrete event simulation. *Proc. of the IEEE*, 77(1):14–23, 1989.

7. J.-P. Katoen. *Quantitative and Qualitative Extensions of Event Structures.* PhD thesis, Universiteit Twente, 1996.

8. J.-P. Katoen, E. Brinksma, D. Latella, and R. Langerak. Stochastic simulation of event structures. In M. Ribaudo, editor, *Proc. of PAPM 1996*, pages 21–40. C.L.U.T. Press, 1996.

9. R. Langerak and H. Brinksma. A complete finite prefix for process algebra. In *CAV'99*, LNCS 1663, pages 184–195. Springer, 1999.

10. K. L. McMillan. *Symbolic Model Checking*, chapter 9: A Partial Order Approach, pages 153–167. Kluwer, 1993.

11. V. Mertsiotakis and M. Silva. A throughput approximation algorithm for decision free processes. In M. Ribaudo, editor, *Proc. of PAPM 1996*, pages 161–178. C.L.U.T. Press, 1996.

12. G. Olsder, J. Resing, R. de Vries, M. Keane, and G. Hooghiemstra. Discrete event systems with stochastic processing times. *IEEE Transactions on Automatic Control*, 35(3):299–302, 1990.

13. T. C. Ruys, R. Langerak, J.-P. Katoen, D. Latella, and M. Massink. First passage time analysis of stochastic process algebra using partial orders. In T. Margaria and W. Yi, editors, *Proc. of TACAS 2001*, LNCS 2031, pages 220–235. Springer, 2001.

Expressing Processes with Different Action Durations through Probabilities

Mario Bravetti and Alessandro Aldini

Università di Bologna, Dipartimento di Scienze dell'Informazione,
Mura Anteo Zamboni 7, 40127 Bologna, Italy,
{bravetti,aldini}@cs.unibo.it

Abstract. We consider a discrete time process algebra capable of (i) modeling processes with different probabilistic advancing speeds (mean number of actions executed per time unit), and (ii) expressing probabilistic external/internal choices and multiway synchronization. We show that, when evaluating steady state based performance measures expressed by associating rewards with actions, such a probabilistic approach provides an exact solution even if advancing speeds are considered not to be probabilistic (i.e. actions of different processes have a different exact duration), without incurring in the state space explosion problem which arises with an intuitive application of a standard synchronous approach. We then present a case study on multi-path routing showing the expressiveness of our calculus and that it makes it particularly easy to produce scalable specifications.

1 Introduction

The modeling experience in the specification of probabilistic concurrent systems (see, e.g., [4] and the references therein) has revealed the importance of using languages expressing advancing speed of processes, probabilistic internal/external choices, and multi-way synchronization for representing the behavior of real systems. In [1] we have considered a probabilistic calculus that combines, in a natural way, these mechanisms. In particular, such a calculus adopts a mixture of the generative and reactive approaches [5] by considering an *asymmetric* form of synchronization where a process which behaves generatively may only synchronize with processes which behave reactively. The integration of the generative and reactive approaches is naturally obtained, similarly as in [12], by designating some actions, called *generative* actions, as predominant over the other ones, called *reactive* actions (denoted by a subscript "*"), and by imposing that generative actions can synchronize with reactive actions only. In particular, the parallel operator that we considered in [1] is similar to the CSP [6] operator $P \parallel_S Q$, where processes P and Q are required to synchronize over actions of type in the set S and locally execute all the other actions. Such an operator expresses multi-way synchronizations by assuming that the result of the synchronization of a generative action a and a reactive action a_* is a generative action a, while the result of the synchronization of two reactive actions a_* is again a reactive action a_*.

L. de Alfaro and S. Gilmore (Eds.): PAPM-PROBMIV 2001, LNCS 2165, pp. 168–183, 2001.

As a consequence, an n-way synchronization is composed of all reactive actions except at most one generative action: the choice of a generative action a determines the action type to be performed and the other processes internally react by *independently* choosing one of their reactive actions a_*. Similarly as in [3], our parallel operator is parameterized with a probability p, determining the process performing the next move in a term $P \|_S^p Q$: we choose P with probability p while we choose Q with probability $1 - p$. We call p and $1 - p$ the *probabilistic advancing speeds* of P and Q, respectively. As an example, let us consider a system composed of two sequential processes whose behavior is described by a term like[1]: $(a +^p b) \|^q (c +^r d)$. According to [3], we first choose which of the two processes must make the next move according to probabilities q and $1 - q$. Then, if the lefthand process wins we locally choose between a and b according to probabilities p and $1 - p$, otherwise if the righthand process wins we locally choose between c and d according to probabilities r and $1 - r$. Moreover, our approach integrates the generative-reactive approach inspired by [12] with the approach to probabilistic process choice inspired by [3] in that the selection of the action to be executed in a system state is conceptually carried out through two steps (a generative choice determining the action type followed by a reactive choice), each possibly employing probabilistic choice among processes. Since we see the reactive actions as *incomplete* actions which must synchronize with generative actions of another system component in order to form a complete system, *fully specified* systems always give rise to probabilistic transition systems which are purely generative. Fully specified systems are therefore fully probabilistic systems (systems not including non-deterministic choices [9]), from which a Markov Chain can be trivially derived by discarding actions from transition labels. As a consequence, they can be easily analyzed to derive performance measures.

In this paper we start by showing that, if we interpret probabilistic transition systems produced with the calculus of [1] in a *discrete time setting*, where each transition takes a discrete time step to be executed, actions of processes are actually executed with the advancing speed expressed as parameters of the parallel operators.

Differently from existing discrete time process algebras, where parallel processes are executed in synchronous locksteps (see, e.g., [7,11,8]), the parallel operator that we adopt is asynchronous and allows processes with different *probabilistic advancing speeds* (mean number of actions executed per time unit) to be modeled. As we now show, $P \|_S^p Q$ represents a system where the mean action frequency (number of actions executed per time unit) of process P is p, while the mean action frequency of process Q is $1 - p$. Since P and Q may advance at different action frequencies, with respect to the classical synchronous approach, modeling a concurrent system does not necessarily imply adopting the same duration for the actions of P and Q. For instance, we could model a post office with a priority mail and an ordinary mail service simply by the term $P \|_\emptyset^{0.2} Q$, where: process P, representing the ordinary mail service, repeatedly executes actions a expressing the delivery of a letter via ordinary mail; process Q, representing the

---

[1] We use $\|^p$ instead of $\|_S^p$ when $S = \emptyset$.

priority mail service, repeatedly executes actions b expressing the delivery of a letter via priority mail. Supposed that we take minutes to be the time unit on which the post office specification is based, in $P \|_\emptyset^{0.2} Q$ the mean frequency for the ordinary mail service is 0.2 letters per minute (288 per day) and the mean frequency for the priority mail service is 0.8 letters per minute (1152 per day). Therefore the actions of P take 5 minutes in the mean to be executed, while the actions of Q take 1 minute and 15 seconds in the mean to be executed.

To be more precise the execution of a system $P \|_S^p Q$ is determined by assuming a probabilistic scheduler that in each global state decides which process between P and Q will perform the next step. In particular P and Q advance in discrete steps and the scheduler decides who is going to perform the next move by tossing an unfair coin which gives "head" with probability p and "tail" with probability $1 - p$. If the coin gives "head" P moves, if the coin gives "tail" Q moves. After a certain number, let us say n, of coin tosses, i.e. after n time units, the mean number of heads that have been extracted (steps P has made) is $n \cdot p$ while the mean number of tails that have been extracted (steps Q has made) is $n \cdot (1 - p)$. Formally such mean values are derived in the following way: $n \cdot p$ is the mean value of a discrete random variable following a binomial distribution with parameters n (number of experiments) and p (probability of success for each experiment). This means that P performs a mean of p steps per time unit and Q performs a mean of $1 - p$ steps per time unit. Hence p is P's probabilistic advancing speed while $1 - p$ is Q's probabilistic advancing speed. Note that in the semantics of $P \|_S^p Q$ we do not express an actual concurrent execution of the actions of P and Q (so that when time passes for an action of P then it passes also for a concurrent action of Q). The behavior of a system $P \|_S^p Q$ is, instead, described at a higher level of abstraction by a model with discrete time steps where there is no actual parallel execution, but only an interleaving of the steps of the two processes, where each step takes the same amount of time (a time unit). To make this more clear, we can interpret the semantics of $P \|_S^p Q$ as originated by a *single-processor machine* executing both processes (P and Q) via a probabilistic scheduler. In this view choices in the global states of a system $P \|_S^p Q$ do not represent "races" between concurrent time delays (as it is usual for continuous time process algebras) but only probabilistic choices that determine which is the process performing the next discrete step. Therefore we do not assume memoryless distributed sojourn times as, e.g., in the continuous time model of [12], but we simply execute a system transition every discrete time unit. In the execution of $P \|_S^p Q$ sometimes we perform a discrete move of P and sometimes we perform a discrete move of Q and what matters, from the performance behavior standpoint, is the frequency with which the actions of a given process are executed: even if we represent the system behavior at a certain level of abstraction where actions are not concurrently executed but just interleaved, we have that such a representation gives the correct execution frequencies for actions of processes. Representing the system behavior just taking care of execution frequencies is the level of abstraction necessary in order to evaluate performance properties in discrete time. This because, since the states of a Discrete Time

Markov Chain (DTMC) are not endowed with different sojourn times (like e.g. in Continuous Time Markov Chains), the evaluation of performance measures in a DTMC is entirely based on the execution frequency of transitions. As we will see, the operator $P \parallel_S^p Q$ can be used both for modeling the execution of P and Q in a single processor machine by means of a probabilistic scheduler (strict interpretation of the semantics of $P \parallel_S^p Q$), and for expressing, as in the post office example, the actual concurrent execution of processes P and Q whose actions have a (possibly) different mean duration (interpretation of the semantics of $P \parallel_S^p Q$ as an abstract description of concurrency), depending on how we calculate the duration of the time unit to be considered for the composed system.

In this paper we also consider the problem of modeling discrete time systems where the different advancing speeds at which concurrent processes proceed are not probabilistic. This means that if, e.g., the action frequency of a process P in a parallel composition is $1/n$, with n natural number, then each action of P takes exactly n time units to be executed. Through a standard approach based on a synchronous parallel composition, where parallel processes are executed in synchronous locksteps (see, e.g., [7,11,8]), processes executing actions with different exact durations could be modeled as follows. Called n_P the duration of the actions of a process P (number of time units taken to execute each action of P), we compute the greater common divisor div of the set of durations n_P, where P is a process composing the system. Then for each process P we split each action in n_P/div subactions and we take div to be the action duration in the specification of the whole system. Such an approach has the problem that the state space of the system greatly increases due to the splitting of the actions of processes. Our approach constitutes an approximated solution to this problem, in that action frequencies of processes are probabilistic instead of being exact, but actions are not split. Nevertheless we show that, in the case of non-blocking processes (i.e. processes enabling at least a generative action in each state) while such an approximation may affect the performance behavior of the system during an initial transient evolution, it gives correct performance measures when the system reaches a steady behavior. Therefore as far as the evaluation of steady state based performance measures of systems is concerned (at least if they are expressible by associating rewards with actions [4]), our approach avoids action splitting, hence the state space explosion problem, while preserving the possibility of exactly analysing concurrent processes with exact advancing speeds.

Finally we present a case study which shows all the main features of our calculus. More precisely, we model and analyze an algebraic specification of a router implementing a probabilistic multi-path routing mechanism. This case study shows that: (i) our approach makes it possible to analyze systems whose components are specified through actions with largely different exact durations (in our model we have actions lasting from half a microsecond to 20 milliseconds); (ii) expressing advancing speeds of processes via a parameterized parallel operator (instead of, e.g., via weights attached to the actions they can perform) is convenient from a modeling viewpoint because the modeler can first specify the behavior of processes in isolation and then establish, independently on how they are specified, their advancing speed when composing them in parallel; (*iii*) thanks to the use of our probabilistic parallel operator and to the generative-reactive mechanism, it is possible to define a specification of the router which is easily scalable to an arbitrary size of the routing table.

The paper is organized as follows. In Sect. 2 we briefly recall the syntax of the calculus of [1]. In Sect. 3 we show how to calculate the time unit to be considered for the composed system depending on the interpretation (processes executed by a single processor machine or actually concurrent processes) given to the parallel operator. In Sect. 4 we show that through our approach we can model systems where actions of different processes have a different exact duration. Finally, in Sect. 5 we present the case study on multi-path routing.

2   Syntax of the Calculus

In this section we briefly recall the generative-reactive calculus introduced in [1]. Formally, we denote the set of action types by *AType*, ranged over by $a, b, \ldots$. As usual *AType* includes the special type τ denoting internal actions. We denote the set of reactive actions by $RAct = \{a_* \mid a \in AType\}$ and the set of generative actions by $GAct = AType$. The set of actions is denoted by $Act = RAct \cup GAct$, ranged over by $\pi, \pi', \ldots$. The syntax of the generative-reactive process algebra is defined as follows. Let *Const* be a set of constants, ranged over by $A, B, \ldots$.

The set $\mathcal{L}$ of process terms is generated by the syntax:

$$P ::= \underline{0} \mid \pi.P \mid P +^p P \mid P \|_S^p P \mid P[a \to b]^p \mid A$$

with $a \in AType - \{\tau\}, b \in AType, S \subseteq AType - \{\tau\}, p \in]0, 1[$. The set $\mathcal{L}$ is ranged over by $P, Q, \ldots$. We denote by $\mathcal{G}$ the set of guarded and closed terms of $\mathcal{L}$.

We use the following abbreviations to denote the relabeling of several action types. Let "$P[\varphi]$", where φ is a finite sequence $\langle (a_1, b_1)^{p_1}, \ldots, (a_n, b_n)^{p_n} \rangle$ of pairs of actions (a_i, b_i) such that $a_i \neq \tau$, with an associated probability p_i, stand for the expression $P[a_1 \to b_1]^{p_1} \ldots [a_n \to b_n]^{p_n}$, relabeling the actions of type $a_1, \ldots, a_n$ into the visible actions $b_1, \ldots, b_n$, respectively. For the sake of simplicity, we omit probabilistic parameters in the operators of our calculus whenever they are not meaningful. For a presentation of the semantics of the calculus we refer to [1,2].

3   Expressing Processes with Different Action Durations

In this section we show how our probabilistic parallel composition operator $P \|_S^p Q$ can be used to express the concurrent execution of processes P and Q whose actions have a different duration. If we strictly follow this single-processor interpretation of the semantics of $P \|_S^p Q$, we assume that the specifications of processes P and Q are based on the same time unit u representing action duration, and consequently that actions of $P \|_S^p Q$ also take time u to be executed, i.e. u is also the time unit of $P \|_S^p Q$. Since P and Q must share a single resource (the processor), the effect of putting P in parallel with Q is that both P and Q

get slowed down. In particular, when P (Q) is considered in isolation it executes one action per time unit u, when, instead, it is assumed to be in parallel with Q (P) by means of $P \|_\emptyset^p Q$, it executes $p(1-p)$ actions per time unit u.

On the other hand such an action interleaving based representation of $P \|_S^p Q$ can be interpreted as being an abstract description of the actual concurrent execution of two processes P and Q specified with respect to (possibly) different action durations, as in the case of the post office example of Sect. 1. In general, if f_P is the mean action frequency assumed in the specification of P (an action of P is assumed to take time $1/f_P$ in the mean to be executed) and f_Q is the mean action frequency assumed for the specification of Q, it is easy to derive a time unit u and a probability p such that, if we assume actions to take time u to be executed, $P \|_\emptyset^p Q$ represents the actual concurrent execution of processes P and Q with mean action frequencies f_P and f_Q, respectively. Since the mean action frequency of process P is f_P and the mean action frequency of process Q is f_Q, the mean action frequency of $P \|_\emptyset^p Q$ must be $f = f_P + f_Q$. Therefore the time unit representing the duration of the actions of $P \|_\emptyset^p Q$ that we have to consider is $u = 1/f = 1/(f_P + f_Q)$, and the action frequency p of P with respect to the new time unit u is given by $f_P = p/u = p \cdot f$, hence $p = f_P/f = f_P/(f_P + f_Q)$. Similarly the action frequency $1-p$ of Q with respect to u turns out to be $1 - p = f_Q/f = f_Q/(f_P + f_Q)$. It is worth noting that by adopting a suitable time unit in this way, the speed at which P and Q are executed when they are considered in isolation is not reduced when they are executed in parallel.

4   Expressing Processes with Exact Advancing Speeds

In this section we show that when evaluating steady state based performance measures, we are able to deal with processes proceeding with different advancing speeds which are not probabilistic. In particular, while during an initial transient evolution considering the action frequency of processes as being exact instead of probabilistic may lead to different results when evaluating performance, in the case of non-blocking processes this does not happen when the system reaches a limiting steady behavior.

As already explained in Sect. 1, executing a process P in parallel with another process which proceeds with a different action frequency, could be done through a standard approach based on a synchronous parallel composition by adequately scaling the time unit on which the specification of P is based, i.e. by splitting each action of P in a certain number n of subactions (see [2] for the details).

Example 1. Let us consider a communication system composed of a processing unit receiving messages from an incoming channel and, after some internal computation, sending out them to an outgoing channel. Suppose that P is a specification of such a system where the time unit is considered to be a second, i.e. an action takes one second to be executed, and $\mathbf{P}$ is the underlying DTMC. Suppose that we want to evaluate the throughput of the system in terms of the number of messages sent out per time unit. In order to do this we consider a reward structure, where we associate a reward equal to 1 to each action representing the sending of a message and a reward equal to 0 to each other action of the system specification. Now, let us suppose that we want to express the behavior of the system with respect to a different time unit, e.g. tenth of seconds instead of seconds. We may need to do this because we want to execute it in parallel with another process whose specification is made in tenth of seconds. Scaling the time unit by a factor $1/10$ can be made by splitting each action a of P in 10 subactions (9 *idle* actions followed by action a), thus obtaining a scaled DTMC $\mathbf{P}'$. The reward structure that we consider for the time scaled system is unchanged with respect to the reward structure considered for the original system. This because the reward gained by actions is not related with their duration but it is just used to count the occurrences of the actions, i.e. rewards 0 and 1 associated to actions are not durations expressed in seconds, but just numbers. By calculating the throughput of the time scaled system (see [2]), we obtain $m_t/10$, i.e. one tenth of the throughput m_t of the original system. This is an expected result because the throughput is a frequency which is expressed in number of actions executed per time unit and we changed the time unit from seconds to tenth of seconds.

Now let us suppose that we want to evaluate the utilization of the processing unit in terms of the percentage of time occupied by the system in performing internal computations for message processing. We consider a reward structure where we associate a reward 1 to each action representing an internal computation of the processing unit and a reward equal to 0 to each other action of the system specification. Now, similarly as in the previous case, let us suppose that we want to scale the time unit by a factor $1/10$. Considered the DTMC $\mathbf{P}'$ obtained by scaling $\mathbf{P}$, we have to evaluate the reward structure to be associated with the time scaled system. Differently from the case of the message throughput, the reward gained by an action is related to the duration of the activity it represents, i.e. rewards 0 and 1 are durations expressed in seconds. Hence, when we consider as the time unit tenth of seconds instead of seconds each reward must be multiplied by 10. By calculating the utilization of the time scaled system (see [2]), we obtain the same utilization m_u of the original system. This is an expected result because the percentage of utilization of the processing unit does not change if we scale the time unit for each activity of the system. □

By employing our probabilistic parallel composition operator we can approximate the scaling of a factor $1/n$ of the time unit used in a specification P by executing P with a probabilistic action frequency $1/n$. This is obtained by considering, e.g., the term $P \parallel_{\emptyset}^{p} Idle$, where $Idle \overset{\Delta}{=} idle.Idle$ and $p = 1/n$.

Theorem 1. *The steady state based performance measures of $P \parallel_{\emptyset}^{p} Idle$ expressible by attaching rewards to the generative actions of P are exactly as those derived by executing the generative actions of P with an exact frequency p.* [2] □

---

[2] In this theorem and in the following Theorem 2 we consider performance measures of periodic DTMCs to be evaluated from their time averaged steady state probabilities. The proof of both theorems can be found in [2].

Example 2. Let us consider the communication system P of Example 1, where the time unit is taken to be one second and $\mathbf{P}$ is the DTMC underlying P. Now we see how to evaluate the two performance measures of Example 1 by employing our approach based on probabilistic parallel composition. We scale the time unit of P to tenth of seconds by executing P with a probabilistic action frequency $1/10$. In particular we consider the term $P \|_{\emptyset}^{1/10}$ *Idle* and its underlying DTMC $\mathbf{P}'$. Both in the case of the system throughput and in the case of the processing unit utilization (by using the same reward structure for the time scaled system as that considered in Example 1), we obtain the same performance measures as with the approach based on action splitting (see [2]).   □

It is worth noting that, with respect to the standard approach based on action splitting, scaling the time unit via our probabilistic parallel operator gives the new possibility of using scaling factors p which are not of the form $p = 1/n$ with n natural number. In general $P \|_{\emptyset}^{p}$ *Idle*, when considered at the steady state, scales the time unit u_P used in the specification of P of a factor p: from $1/u_P = p/u$ we derive $u = u_P \cdot p$ where u represents action duration in the behavior of $P \|_{\emptyset}^{p}$ *Idle*.

Finally, we have that with our probabilistic parallel operator we obtain a correct time unit scaling at the steady state also for non-blocking processes which are part of a larger system. Hence we can express the parallel composition of processes specified with respect to different time units through a common time unit by considering a different scaling factor for the time unit of each process.

Theorem 2. *Supposed that both P and Q never block during the execution of $P \|_{S}^{p} Q$ (in all the states of $P \|_{S}^{p} Q$ at least one generative action of P and one generative action of Q are executable), the steady state based performance measures of $P \|_{S}^{p} Q$ expressible by attaching rewards to the generative actions of P and Q are axactly as those derived by executing the generative actions of P and Q with an exact frequency p and $1 - p$, respectively.*   □

Therefore given two different time units u_P and u_Q representing action duration in the specifications of P and Q in isolation, respectively, then there exists a common time unit u and a probability p (determined from $f_P = 1/u_P$ and $f_Q = 1/u_Q$ as explained in Sect. 3) such that $P \|_{S}^{p} Q$ just expresses the concurrent execution of the two processes without affecting their advancing speed.

5   A Case Study: Multi-path Routing

In this section we present a case study showing how our approach provides a compositional and intuitive method for specifying concurrent systems in a scalable way. In particular, we consider a multi-path routing mechanism of the OSI network layer [10], and we model and analyze an internetworking node (termed Interface Message Processor, IMP for short), whose arriving packets have several possible destinations with several possible ways to reach a destination.

The routing algorithm decides, at the network layer, on which output link an arriving packet should be sent, depending on the destination of that packet.

We abstract from the algorithm used to determine an optimal path between two nodes of a network, and we assume that the modeled IMP has a routing table including the route information with several possible choices for each destination. A weight is associated to each possible path and these weights are used as probabilities to decide where to send the present packet. Supporting multiple paths to the same destination, unlike single-path algorithms, permits traffic multiplexing over multiple lines and substantially provides better throughput and reliability.

5.1   Algebraic Specification of the Multi-path Router

The overall model of our IMP (term *Multipath*) is shown in Table 1 in the case of two possible destinations (a and b) and two possible paths for each destination (a_1, a_2 for a, and b_1, b_2 for b). The algebraic specification is composed of several processes which are actually concurrent and are specified with respect to different time units. In particular, system *Multipath* consists of three concurrent components: a term *Arrivals* modeling the incoming traffic, a term *Router* modeling the core of the IMP, and a term *Channels* modeling the outgoing channels. The structure of the three components is as follows: (*i*) Term *Arrivals* is composed of two concurrent processes *Arrivala* and *Arrivalb* which model the incoming traffic directed to destinations a and b, respectively. The time unit representing action duration that we consider for both processes is one millisecond. The adoption of such a time unit makes it easy to represent a realistic workload for the IMP. In particular we assume that for each destination at most one packet per millisecond can arrive to the IMP, i.e. the maximum frequency of the incoming traffic is 2000 packets per second. (*ii*) Term *Router* represents a process whose time unit is half a microsecond, i.e. it can execute 2000000 actions per second. As we will see, the *Router* term, which is the core of the IMP, is a single processor machine managing the packets directed to the two destinations via a probabilistic scheduler. (*iii*) Term *Channels* is composed of four concurrent processes modeling the possible channels a_1, a_2, b_1, and b_2. Since we take packet transmission to be represented by the execution of a corresponding action, their time units are defined on the basis of their bandwidth. The time unit for the channel a_1 is 10 milliseconds, i.e. it can send out 100 packets per second, while the time unit for the channel a_2 is 2.5 milliseconds (400 packets per second). The time units for the channels b_1 and b_2 are 5 milliseconds (200 packets per second) and 4 milliseconds (250 packets per second), respectively.

Note that it is possible to compose in parallel the above processes, which are specified with respect to different time units, because, as it can be easily verified, each of them never blocks during system execution. In order to express the actual concurrent execution of such processes, all the time units used in their specification are scaled to a common global time unit u. More precisely, u is evaluated by computing the inverse of the global action frequency of the composed system. Hence, in our case study u is the inverse of $1000+1000+2000000+100+400+200+250$ actions per second, i.e. $u = 1/2002950$ seconds. Such a global time unit u determines the action frequencies to be considered as parameters of the parallel operators used to describe the concurrent ex-

Table 1. Multi-path Routing Model

$$
\begin{aligned}
Multipath &\stackrel{\Delta}{=} Arrivals \parallel_S^{p_1} (Router \parallel_C^{p_2} Channels) \\
S &= \{receivea,\ receiveb\} \\
C &= \{avail\_cha_1,\ avail\_cha_2,\ avail\_chb_1, \\
&\qquad avail\_chb_2,\ transma_1,\ transma_2,\ transmb_1,\ transmb_2\} \\[4pt]
Arrivals &\stackrel{\Delta}{=} Arrivala \parallel^{\frac{1}{2}} Arrivalb \\
Channels &\stackrel{\Delta}{=} (Channel[\varphi'_1] \parallel^{pa} Channel[\varphi'_2]) \parallel^{v} (Channel[\varphi''_1] \parallel^{pb} Channel[\varphi''_2]) \\
Router &\stackrel{\Delta}{=} (Queues \parallel_{S_1 \cup I} Switch) \parallel_I Idle \\
S_1 &= \{accepta,\ acceptb\} \qquad I = \{idle\} \\[4pt]
Queues &\stackrel{\Delta}{=} Queue[\varphi'] \parallel_I Queue[\varphi''] \\
Switch &\stackrel{\Delta}{=} (Manager[\varphi'] \parallel_{S'_2} (Routing[\varphi'][\varphi'_1] \parallel_{B'}^{qa} Routing[\varphi'][\varphi'_2]))\ \parallel_I^{p} \\
&\qquad (Manager[\varphi''] \parallel_{S''_2} (Routing[\varphi''][\varphi''_1] \parallel_{B''}^{qb} Routing[\varphi''][\varphi''_2])) \\
S'_2 &= \{senda,\ busya\} \qquad S''_2 = \{sendb,\ busyb\} \\
B' &= \{busya\} \qquad B'' = \{busyb\} \\
\varphi' &= \langle(receive, receivea),\ (accept, accepta),\ (send, senda),\ (busy, busya)\rangle \\
\varphi'' &= \langle(receive, receiveb),\ (accept, acceptb),\ (send, sendb),\ (busy, busyb)\rangle \\
\varphi'_i &= \langle(transm, transma_i),\ (avail\_ch, avail\_cha_i)\rangle\ i \in \{1,2\} \\
\varphi''_i &= \langle(transm, transmb_i),\ (avail\_ch, avail\_chb_i)\rangle\ i \in \{1,2\} \\[4pt]
Arrivala &\stackrel{\Delta}{=} receivea.Arrivala +^{ra} wait.Arrivala \\
Arrivalb &\stackrel{\Delta}{=} receiveb.Arrivalb +^{rb} wait.Arrivalb \\
Queue &\stackrel{\Delta}{=} receive_*.Queue' + idle_*.Queue \\
Queue' &\stackrel{\Delta}{=} receive_*.accept_*.Queue' + accept_*.Queue \\
Manager &\stackrel{\Delta}{=} accept.Manager' + idle_*.Manager \\
Manager' &\stackrel{\Delta}{=} send.Manager + busy.Manager' \\
Routing &\stackrel{\Delta}{=} send_*.Routing' + avail\_ch_*.Routing \\
Routing' &\stackrel{\Delta}{=} transm_*.Routing + busy_*.Routing' \\
Idle &\stackrel{\Delta}{=} idle.Idle \\
Channel &\stackrel{\Delta}{=} avail\_ch.Channel + transm.Channel
\end{aligned}
$$

ecution of system components in Table 1. In particular, parameter p_1 representing the advancing speed of term *Arrivals* in $Arrivals \parallel_S^{p_1} (Router \parallel_C^{p_2} Channels)$ is given by the ratio of the action frequency of term *Arrivals* over the global action frequency of system *Multipath*, i.e. if we express action frequency in seconds $p_1 = 2000/2002950 \approx 0.000999$. Similarly, parameter p_2 representing the advancing speed of term *Router* in $Router \parallel_C^{p_2} Channels$ is given by the ratio of the action frequency of term *Router* over the global action frequency of term $Router \parallel_C^{p_2} Channels$, i.e. $p_2 = 2000000/(2000000 + 950) \approx 0.999525$. As far as the specification of the *Arrivals* component is concerned, the parallel composition of the two concurrent processes *Arrivala* and *Arrivalb* has parameter $\frac{1}{2}$ because their action frequency is the same (1000 actions per second). As far as the specification of the *Channels* component is concerned, in $(Channel[\varphi'_1] \parallel^{pa} Channel[\varphi'_2]) \parallel^{v} (Channel[\varphi''_1] \parallel^{pb} Channel[\varphi''_2])$ we take: pa to be $100/(100 + 400) = 0.2$; pb to be $200/(200 + 250) \approx 0.444444$; and v to be

$500/(500 + 450) \approx 0.526316$. As an example of the calculated action frequencies, process *Arrivala* executes $p_1 \cdot \frac{1}{2} \approx 0.000499$ actions per time unit u, i.e. 1000 actions per second, and process *Router* executes $(1 - p_1) \cdot p_2 \approx 0.998527$ actions per time unit u, i.e. 2000000 actions per second.

Now let us describe in detail the behavior of each process of the system: (i) Process *Arrivala* (*Arrivalb*) models the incoming traffic through a Bernoulli distribution with parameter ra (rb). In particular, an arriving packet is represented by the action *receivea* (*receiveb*) which synchronizes with the corresponding reactive action in the queue for packets a (b) of term *Router*. In the case such a queue is full the action *receivea* (*receiveb*) is not enabled and the arriving packets are lost (the generative action *wait* is executed with probability 1). (ii) Process *Router* is the core of the IMP and is composed of a term *Queues* collecting the arriving packets, a term *Switch* delivering the packets to the outgoing channels, and a term *Idle* modeling the phases of router inactivity. They are defined as follows. Term *Queues* consists of two *Queue* processes, one for each kind of packet, which behave reactively. In particular, they receive packets destined to a (b) through reactive actions of type *receivea* (*receiveb*) and pass them to the *Switch* term through reactive actions of type *accepta* (*acceptb*). For the sake of simplicity we assume both queues to be of size 2. Term *Switch* is a single-processor machine executing two different terms, each one managing packets with a certain destination (a or b), via a probabilistic scheduler with parameter p. In this way, by varying p, we can model an IMP that delivers packets with a particular destination more efficiently than packets with another destination, e.g. for commercial reasons. The term delivering packets to destination a (b) is composed of a *Manager* and two *Routing* terms, each one delivering packets to a particular channel a_1 or a_2 (b_1 or b_2). Term *Manager* accepts packets destined to a (b) from the dedicated queue through action *accepta* (*acceptb*) and afterwards either immediately passes them to one of the two *Routing* terms through action *senda* (*sendb*), or waits until at least one channel is available for transmission by performing action *busya* (*busyb*). This behavior is realized through a generative-reactive mechanism as follows. The two *Routing* terms behave reactively and each of them accepts packets through a reactive action of type *senda* (*sendb*) and transmits them through the corresponding channel via a reactive action of type *transma* (*transmb*). Whenever the generative action *senda* (*sendb*) is enabled by the *Manager* term, the *Routing* term accepting the packet is chosen according to the probability qa (qb) parameterizing the parallel composition of the two *Routing* terms. Note that a *Routing* term may be not available for accepting a packet because it is currently transmitting through action *transma* (*transmb*) a packet previously received. Therefore, whenever only one *Routing* process is available for accepting a packet coming from the *Manager* term, the packet is transmitted through the corresponding channel with probability 1. Whenever both *Routing* processes are busy, the transmission of packets destined to a (b) is not possible and this is signalled to the *Manager* term through a multiway synchronization by enabling the reactive action of type *busya* (*busyb*). Term *Idle* executes an action *idle* (representing the fact that the IMP is idle) whenever term *Router* has nothing else to do. More precisely, action *idle* is executed through a multiway synchronization with all the other *Router* components if and only if the input queues (term *Queues*) are empty and the core of the IMP (term *Switch*) is not waiting for delivering a packet to the channel. In particular, term *Idle* prevents the term *Router* from blocking, thus allowing the advancing speed of terms *Arrivals*, *Router*, and *Channels* to be preserved and satisfying the condition needed for composing processes with different time units. (*iii*) The four *Channel* processes model the outgoing channels a_1, a_2, b_1 and b_2. Each process *Channel* can either be transmitting a packet when the generative action *transm* is synchronized with the corresponding reactive action of term *Routing* managing that channel, or be available for transmission when the generative action *avail_ch* is synchronized with the corresponding reactive action of term *Routing*. For instance in the case of channel a_1 the generative actions $transm a_1$ and $avail\_cha_1$ must synchronize with the reactive actions $transm a_{1_*}$ and $avail\_cha_{1_*}$ of the *Routing* term managing channel a_1, respectively. In this way the generative actions of a *Channel* process are executed in mutual exclusion in the sense that in every system state one and only one of them is enabled. As a consequence term *Channels* never blocks.

Thanks to our approach which allows processes specified with respect to different time units to be modeled without splitting actions, we have that the transition system underlying the algebraic specification of Table 1 is composed of 576 states and 4768 transitions only. This is a crucial result, because if we model the same system by resorting to a classical approach which scales the time unit by splitting each action, we have to cope with the serious problem of a greatly increased size of the state space. For instance, since the basic time unit for the router is half a microsecond while the basic time unit for the input channels is a millisecond, in order to compose in parallel terms *Arrivals* and *Router* we have to split the actions of term *Arrivals* in thousands of subactions thus causing a state space explosion. Moreover, we point out that the generative-reactive behavior of the *Switch* process represents the core of this case study. In particular, process *Switch* generatively decides, according to probability p, which of the two *Manager* terms performs a send action (*senda* for the manager delivering packets to destination a or *sendb* for the other one), while it reactively decides, according to probability qa or qb (depending on the *send* action performed) which of the terms *Routing* synchronizes with such an action. A calculus capable of expressing generative-reactive choices is, therefore, very suitable (if not necessary) to model systems with such a behavior. Finally, it is worth noting that, thanks to the choice of putting probabilities in the operators instead of, e.g., attaching them to actions, and to the expressive power of our generative-reactive approach, it was possible to specify the IMP in such a way that all the probabilistic mechanisms on which its behavior is based (and which are not related with the internal behavior of a process) depend on the parameters of parallel composition operators only. As a consequence, scaling the system specification to a higher number of components does not make it necessary to change the internal behavior of processes. For instance, in [2] we have scaled the system of the router specification to four possible channels for each destination, by simply adding several instances of *Routing* and *Channel* terms, and by appropriately adjusting the parameters of parallel composition operators.

5.2   Performance Analysis

In order to derive performance measures from the multi-path router specification, we resorted to the software tool TwoTowers [4], that has been recently extended to support the generative-reactive approach presented in this paper. Such a tool also implements an algebraic reward based method to specify and derive performance measures. The results of our performance analysis are shown in Fig. 1-3. In particular, we concentrated on two main metrics. On the one hand, we evaluate the throughput of the system at steady state, represented by occurrences of actions of type $transma_1$, $transma_2$, $transmb_1$, and $transmb_2$, by attaching a reward equal to 1 to the above actions and a reward equal to 0 to each other action. Since the throughput is a frequency expressed in terms of number of actions executed per time unit and the time unit is $1/2002950$ seconds, we have to multiply the throughput resulting from the analysis of the Markov Chain by 2002950 in order to obtain the results (expressed in seconds) shown in our tables [2]. On the other hand, we evaluate the router idleness at steady state in terms of the percentage of time the IMP is inactive. The router is considered to be idle when no packet is currently inside the IMP, i.e. when it executes actions of type *idle*. Therefore we attach a reward equal to 1 to such actions and a reward equal to 0 to each other action. Since the time unit of the *Router* process (half a microsecond) is scaled by a factor $(1 - p_1) \cdot p_2 \approx 0.998527$ and the reward gained by actions is related to the duration of the corresponding activity expressed in half microseconds, due to the time unit change we must multiply each reward by $1/0.998527$ before analyzing the Markov Chain [2].

For each conducted analysis, we assumed that the incoming traffic for the destinations a and b follows the same Bernoulli distribution of parameter $r = ra = rb$. The figures show how the performance measures change when we vary r from 0.1 (sometimes 0.01) to 0.9. In this way we observe the system behavior under various levels of workload ranging from 10% (or 1%) to 90%.

We start by evaluating the system throughput under different circumstances. We first consider the situation in which $p = \frac{1}{2}$, i.e. the packets destined to a and b are managed at the same speed by the *Switch* process, and parameters qa and qb reflect the bandwidth distribution over channels directed to destinations a and b, respectively. Since channel a_1 can deliver 100 packets per second and channel a_2 can deliver 400 packets per second, we take $qa = 100/(100 + 400) = 1/5$ (the ratio of the bandwidth of channel a_1 over the overall bandwidth of the channels directed to a), so that packets are probabilistically distributed between channels a_1 and a_2 in the optimal way. Similarly, since channel b_1 can deliver 200 packets per second and channel b_2 can deliver 250 packets per second, we take $qb = 200/(200 + 250) = 4/9$. The obtained results are reported in Fig. 1. As we can see in Fig. 1(B) the curve representing the total system throughput is characterized by a high slope in correspondence of a low workload and a

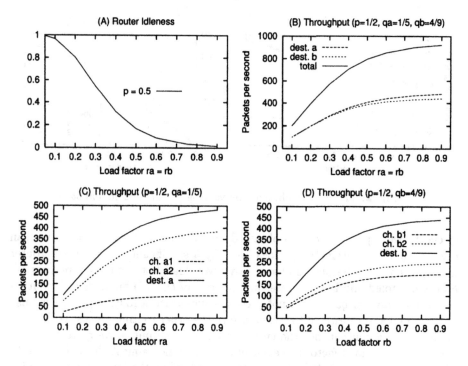

Fig. 1. Idleness and Throughput of the Multi-path Router

quite flat slope when the load factor increases over the 50%. This because, for packets with a given destination, the bandwidth associated with the channels directed to that destination is about one half of the maximum bandwidth of the incoming traffic. Simply put, when the parameter r of the Bernoulli distribution representing the incoming traffic reaches the 50%, the outgoing channel is almost fully occupied, hence a further increment of r gives rise to a very small increment of the outgoing throughput. Another expected result is that the throughput of packets destined to a is slightly greater than the throughput of packets destined to b. This because the overall bandwidth of the channels directed to a is 500 packets per second, while it is 450 packets per second for the channels directed to b. Fig. 1(C) and 1(D) report the throughput for each single channel a_1, a_2, b_1 and b_2. In the case of a_1 and a_2 the distance between the two curves is quite great, this because a_1 has just one fourth of the bandwidth of a_2. As expected such a difference is smaller in the case of b_1 and b_2, because their bandwidth is quite similar (200 and 250 packets per second, respectively).

Since in a realistic framework the value of parameters qa and qb are established by the multi-path routing algorithm governing the IMP according to the network conditions (e.g. estimated time for a packet to reach a destination via a particular path), we study the effect on the throughput of adopting parameters qa and qb which do not reflect the bandwidth distribution over the outgoing channels. The results of such an analysis are reported in Fig. 2. In particular, Fig. 2(A) shows how the throughput of a_1 and a_2 varies when changing the value of qa from $\frac{1}{5}$ to $\frac{4}{5}$, i.e. by exchanging the value of qa and $1 - qa$. For the sake

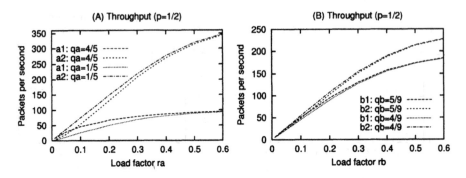

Fig. 2. Throughput obtained by varying parameters qa and qb

of clarity we report the curves obtained for both $qa = \frac{1}{5}$ and $qa = \frac{4}{5}$. We can observe that the parameter qa does not play a significant role when the router is congested. This because under a heavy workload both routing processes are hardly occupied and in most cases at least one of them is busy. In such a situation the parameter qa is often not used, because when a routing process is busy an arriving packet destined to a is passed to the other routing process with probability 1. As a consequence the curves of the throughput converge to the same values when the load factor ra gets over 50%, i.e. when almost all the bandwidth of each channel is exploited. On the other hand, when the incoming workload is low, the parameter qa becomes important as it probabilistically decides which routing process will deliver the packet, hence increasing the throughput of a routing process with respect to the other one (see the quite evident difference among the curves when ra gets under the 30%). Fig. 2(B) shows how the throughput of the two channels destined to b varies when exchanging the value of qb and $1-qb$. With respect to the channels destined to a, in this case the difference between the old value of qb ($\frac{4}{9}$) and its new value ($\frac{5}{9}$) is smaller. This is reflected on the results presented in Fig. 2, where the curves for the old and the new value of qb are almost overlapped for each value of rb.

Now we show the role played by parameter p on the system throughput. In order to merely concentrate on the effects of varying parameter p, we just consider the situation in which parameters qa and qb reflect the bandwidth distribution over channels directed to destinations a and b, respectively. Parameter p can be chosen in order to favour the internal computations of the IMP dedicated to packets destined to a (b) with respect to those dedicated to packets destined to b (a). To this aim, in Fig. 3 we report the throughput of the multi-path router in the case $p = 0.999$, hence when packets destined to b are managed by the IMP much more slowly than packets destined to a. As a consequence of the unfair behavior of the router, we have that the IMP delivers the packets destined to a at the usual speed (the curve for packets destined to a in Fig. 3(A) is the same as that in Fig. 1(B)), but it delays the packets destined to b, hence compromising the throughput of such packets. Therefore with respect to the case $p = 0.5$ the overall system throughput decreases (for easy of comparison in Fig. 3(A) we also report the curve obtained in the case $p = 0.5$). The comparison with the case

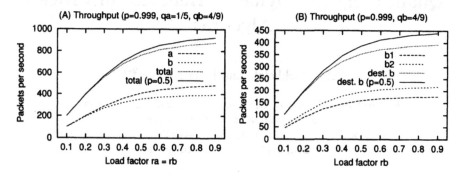

Fig. 3. Throughput obtained by varying parameter p

$p = 0.5$ is even more evident in Fig. 3(B), where we report the throughput of the outgoing channels directed to destination b.

As far as the idleness of the router is concerned, we simply consider the situation in which $p = \frac{1}{2}$ and parameters qa and qb reflect the bandwidth distribution over channels directed to destinations a and b, respectively. The curve presented in Fig. 1(A) shows the relation among the inactivity of the router and the load factor for the incoming traffic. As expected, the router is almost always idle if the workload is low, but the duration of its inactivity phases rapidly converges to zero for a load factor greater than 40%.

References

1. A. Aldini, M. Bravetti, *"An Asynchronous Calculus for Generative-Reactive Probabilistic Systems"*, in Proc. of 8th Int. Workshop on Process Algebra and Performance Modeling, pp. 591-605, 2000
2. M. Bravetti, A. Aldini, *"An Asynchronous Calculus for Generative-Reactive Probabilistic Systems"*, Tech. Rep. UBLCS-2000-03, Univ. Bologna, Italy, 2000
3. J.C.M. Baeten, J.A. Bergstra, S.A. Smolka, *"Axiomatizing Probabilistic Processes: ACP with Generative Probabilities"*, in Inf. and Comp. 121:234-255, 1995
4. M. Bernardo, *"Theory and Application of Extended Markovian Process Algebra"*, Ph. D. Thesis, Univ. Bologna, Italy, 1999
5. R.J. van Glabbeek, S.A. Smolka, B. Steffen, *"Reactive, Generative and Stratified Models of Probabilistic Processes"*, in Inf. and Comp. 121:59-80, 1995
6. C.A.R. Hoare, *"Communicating Sequential Processes"*, Prentice Hall, 1985
7. C.-C. Jou, S.A. Smolka, *"Equivalences, Congruences, and Complete Axiomatizations for Probabilistic Processes"*, in Proc. of 1st Int. Conf. on Concurrency Theory, LNCS 458:367-383, 1990
8. K.G. Larsen, A. Skou, *"Compositional Verification of Probabilistic Processes"*, in Proc. of 3rd Int. Conf. on Concurrency Theory, LNCS 630:456-471, 1992
9. R. Segala, *"Modeling and Verification of Randomized Distributed Real-Time Systems"*, Ph. D. Thesis, MIT, Boston (MA), 1995
10. A. S. Tanenbaum, *"Computer Networks"*, Prentice Hall, 1996
11. C. Tofts, *"Processes with Probabilities, Priority and Time"*, in Formal Aspects of Computing 6:536-564, 1994
12. S. H. Wu, S.A. Smolka, E.W. Stark, *"Composition and Behaviors of Probabilistic I/O Automata"*, in Theoretical Computer Science 176:1-38, 1997

Quantifying the Dynamic Behavior of Process Algebras

Peter Buchholz[1]* and Peter Kemper[2]**

[1] Fakultät für Informatik, TU Dresden, D-01062 Dresden, Germany
p.buchholz@inf.tu-dresden.de
[2] Informatik IV, Universität Dortmund, D-44221 Dortmund, Germany
kemper@ls4.cs.uni-dortmund.de

Abstract. The paper introduces a new approach to define process algebras with quantified transitions. A mathematical model is introduced which allows the definition of various classes of process algebras including the well known models of untimed, probabilistic and stochastic process algebras. For this general mathematical model a bisimulation equivalence is defined and it is shown that the equivalence is a congruence according to the operations of the algebra. By means of some examples it is shown that the proposed approach allows the definition of new classes of process algebras like process algebras over the max/plus or min/plus semirings.

Keywords: process algebras, semiring, bisimulation, congruence.

1 Introduction

Process algebras are one of the most important formal specification techniques to describe the dynamic behavior of discrete event systems. A large variety of different process algebras exists. Process algebras differ in various aspects, e.g., in the way how synchronization is defined, in the used equivalence relations and, may be most important, in the way how dynamics are quantified. First process algebras like Milners CCS [23] or Hoares CSP [21] describe only the functional behavior of systems. Thus, the process algebra specifies which action sequences can potentially occur, but it does not quantify these sequences or relate a notation of time to it. Consequently, process algebras have been extended by relating probabilities to transitions [25,22], by assuming a stochastic timing and assigning rates to transitions [5,8,19,15] and by combining probabilities and time in one model [14]. Also for probabilistic and stochastic process algebras different concrete realizations have been defined which differ in several details. However, almost all process algebras have the following features in common. They have a well defined formal syntax and a semantics defined as some form of a labeled transition system. The specification is compositional. Finally, equivalence

---

* This research is partially supported by DFG, SFB 358
** This research is partially supported by DFG, SFB 559

L. de Alfaro and S. Gilmore (Eds.): PAPM-PROBMIV 2001, LNCS 2165, pp. 184–199, 2001.
© Springer-Verlag Berlin Heidelberg 2001

relations exist which allow to compare different specifications and certain equivalences are congruences according to the operations of the algebra. In summary, these properties make this model type successful.

Known process algebras describe either labeled transition systems or stochastic processes with transition labels. In the latter case mainly Markov processes or Markov chains are considered. However, recently other algebraic models became very popular for the analysis of discrete event systems, namely min/plus, max/plus or min/max systems [2] with applications in the area of communication protocols [4,10], queueing networks [1] or real time systems [3]. Instead of defining additional specific process algebras for these models, it is more challenging to ask for a common framework which has such process algebras a special instantiations. In this paper we follow the latter approach. We first have to find a general representation of the required operations which can be interpreted adequately in different concrete realizations. It will be shown that the mathematical structure for these operations is a semiring. This structure is very general and contains the mathematical operations available in different process algebras and also in systems like max/plus algebra. Based on the semiring definition a general process algebra can be defined by including the common operations available in different process algebras like prefix, sum and composition. The underlying semantics is a labeled transition system where labels consist of actions and an additional quantification by elements of the semiring. The interesting point is that the mild requirements of a semiring as an algebraic structure are sufficient to define a bisimulation in this framework, which appears natural and which is a congruence according to the operations of the algebra. This bisimulation is valid independently of the semiring one selects for a concrete application.

The outline of the paper is as follows. In the next section we introduce some basic definitions. Afterwards, in Sect. 3 we present the syntax and an operational semantics of Generalized Process Algebra (GPA). Then a bisimulation is defined and it is proved that this bisimulation is a congruence according to the operations of the algebra. In Sect. 5 we present two concrete realizations of GPA and define implicitly process algebras for max/plus and min/plus by means of two examples. The paper ends with the conclusions which outline subjects for future research.

2   Basic Definitions

Before we present our process algebra, the underlying basic structures are defined. We start with semirings which are later used to define quantitative values of labels.

Definition 1. *A semiring* $(\mathbb{K}, \widehat{+}, \widehat{\cdot}, \mathbb{0}, \mathbb{1})$ *is a set* $\mathbb{K}$ *with binary operations* $\widehat{+}$ *and* $\widehat{\cdot}$ *defined on* $\mathbb{K}$ *such that the following axioms are satisfied:* $\widehat{+}, \widehat{\cdot}$ *are associative, and* $\widehat{+}$ *is commutative, right and left distributive laws hold for* $\widehat{+}$ *and* $\widehat{\cdot}$, $\mathbb{0}$ *and* $\mathbb{1}$ *are the additive and multiplicative identities with* $\mathbb{0} \neq \mathbb{1}$, *and for all* $k \in \mathbb{K}$ $k \widehat{\cdot} \mathbb{0} = \mathbb{0} \widehat{\cdot} k = \mathbb{0}$ *holds.*

To make the notation simpler we use sometimes $\mathbb{K}$ for the whole semiring.

Some typical examples for semirings are $(\mathbb{B}, \vee, \wedge, 0, 1)$ the Boolean semiring, $(\mathbb{R}, +, \cdot, 0, 1)$ the semiring over the real numbers with the usual addition and multiplication, $(\mathbb{R} \cup \{\infty\}, \min, +, \infty, 0)$ the min/+ semiring, $(\mathbb{R} \cup \{-\infty\}, \max, +, \infty, 0)$ the max/+ semiring or $(\mathbb{R} \cup \{\infty\} \cup \{-\infty\}, \max, \min, \infty, 0)$ the max/min semiring. In the sequel we use for the addition $\widehat{+}$ and for multiplication $\widehat{\cdot}$ or no operator symbol as usual in most semirings.

Semiring operations can be easily extended to define operations on vectors and matrices. Let $\mathbf{p} \in \mathbb{K}^n$ be a n-dimensional vector and $\mathbf{Q} \in \mathbb{K}^{n,n}$ a $n \times n$ square matrix both including elements from semiring $\mathbb{K}$, then $\mathbf{q} = \mathbf{p}\mathbf{Q}$ equals a vector in $\mathbb{K}^n$ which is defined as

$$\mathbf{q}(j) = \widehat{\sum}_{i=0}^{n-1} \mathbf{p}(i)\mathbf{Q}(i,j) \ .$$

where $\widehat{\sum}_{i=0}^{n-1} a_i = a_0 \widehat{+} \ldots \widehat{+} a_{n-1}$. Dot product of vectors, matrix addition and multiplication can be defined similarly.

Now we can define transition systems where transitions are labeled with symbols from a finite alphabet and from a semiring. This notion is subsequently used to define the semantics of our process algebra.

Definition 2. *A (finite) multi labeled transition system (MLTS) is a 5 tuple* $MLTS = (S, Act, \mathbb{K}, T, ini)$, *where S is the state space which is countable (finite), Act is a finite set of transition labels, $\mathbb{K}$ is a semiring used for the definition of transition costs, $T : S \times Act \times S \to \mathbb{K}$ is the transition function, $ini : S \to \mathbb{K}$ is the initialization function.*

An $MLTS$ can be interpreted as a graph where each transition is labeled by a symbol to define the operation and a cost to perform the transition. The meaning of costs can be very different depending on the semiring one considers. If the state space is finite, the $MLTS$ corresponds to an automaton with transition costs or weighted automata, a well known model in automata theory. Each $MLTS$ can be characterized by a set of matrices, one for each $a \in Act$ and a vector. Let $\mathbf{Q}_a$ be the matrix for label $a \in Act$, then $\mathbf{Q}_a(x,y) = T(x,a,y)$. $\mathbf{p}^0$ is the initial vector of the system such that $\mathbf{p}^0(x) = ini(x)$. Function ini gives an initial quantification for all states, which depends on the concrete application, e.g. to consider reachability from a certain state $s_i \in S$ a common definition of ini is $ini(s) = \mathbb{1}$ if $s = s_i$ and $\mathbb{0}$ otherwise. The set of initial states S_{ini} contains all $s \in S$ with $ini(s) \neq \mathbb{0}$.

For many, but not for all, applications, the dynamics of the system can be described by vector matrix products. We briefly describe this kind of dynamics here. Usually one can distinguish between generative and reactive models [25]. In the reactive case, the system reacts to the behavior of some environment which determines the label that occurs next. A state of the system is described by a vector with elements from $\mathbb{K}$. $\mathbf{p}^0$ is the initial state vector and if $\mathbf{p}^k$ describes the current state and $a \in Act$ is the label chosen by the environment, then $\mathbf{p}^{k+1} = \mathbf{p}^k \mathbf{Q}_a$ describes the next state. In a generative situation, the system can decide by itself which transition occurs next. In this situation the next state is computed from $\mathbf{p}^{k+1} = \mathbf{p}^k \widehat{\sum}_{a \in Act} \mathbf{Q}_a$. In both cases we may analyze the system according to the state reached after a transition sequence. Thus, let $\omega \in Act^*$, ω_i be the i-th symbol in ω and $|\omega|$ be the length of ω. $\mathbf{p}^\omega$ is the state reached after observing ω starting with $\mathbf{p}^0$. The vector $\mathbf{p}^\omega$ can be computed as

$$\mathbf{p}^\omega = \mathbf{p}^0 \widehat{\prod}_{i=1}^{|\omega|} \mathbf{Q}_{\omega_i} . \tag{1}$$

where $\widehat{\prod}_{i=0}^{n-1} a_i = a_0 \,\widehat{\cdot}\, \ldots \,\widehat{\cdot}\, a_{n-1}$. We may go further in analyzing the behavior of a system by analyzing its generation of actions (traces) or its reaction to the input from the environment. Define $c_\omega = \widehat{\sum}_{s \in S} \mathbf{p}^\omega(s)$ as the cost of transition sequence ω. Observe that costs can have different meanings in each concrete realization.

Examples: If we use $(\mathbb{B}, \vee, \wedge, 0, 1)$ in the above definition and ini assigns 1 to the initial state and 0 to all other states, then the resulting model describes the well known labeled transition system. Vector $\mathbf{p}^k$ describes the set of reachable states after k transitions. Thus, $\mathbf{p}^k(x) = 1$ implies that x is reachable after k transitions and in the reactive case, the observed labels have been defined by the environment. For deterministic systems, each matrix row contains one element equal to 1.

In a similar way, probabilistic systems can be defined. For probabilistic systems, $\mathbf{p}^0$ has to define a probability distribution. Depending on the used semantics, matrices $\mathbf{Q}_a$ are stochastic or substochastic matrices. In the reactive case, matrices $\mathbf{Q}_a$ are stochastic, since the environment selects label a. In the generative case, each $\mathbf{Q}_a$ is a substochastic matrix and $\sum_{a \in Act} \mathbf{Q}_a$ is a stochastic matrix.

For stochastic systems describing continuous time Markov chains (CTMCs) with transition labels [8,15,20] the above method of computing the dynamic behavior stepwise is not directly applicable. However, by applying the well known randomization technique [13], the CTMC with transition labels can be transformed into a discrete time Markov chain (DTMC) with transition labels and a Poisson process. For the DTMC with transition labels vectors $\mathbf{p}^k$ can be computed and the Poisson process determines the number of transitions which occur in any finite interval.

Another, less common example is the use of max/+ as semiring and interpret the transition labels as reward. For this model element $\mathbf{p}^k(x)$ describes the maximal reward gained when reaching state x after k steps starting in some state y with reward $\mathbf{p}^0(y)$. In the reactive case, the labels of the path from y to x are given by the environment.

For the min/+ semiring and interpretation of transition labels as costs, we have a very similar interpretation. Now $\mathbf{p}^k(x)$ contains the minimal costs of reaching state x after k transitions.

3   A General Process Algebra with Transitions Costs

We now define a process algebra denoted as General Process Algebra (GPA) which uses the above concepts. First the syntax of the process algebra is defined, afterwards the semantics is introduced. The syntax definition follows the usual concepts used in process algebras [23,21].

Definition 3. *The set $\mathcal{L}$ of terms in Generalized Process Algebra (GPA) over a set of finite transition labels Act including label τ and a semiring $\mathbb{K}$ is defined by $\mathcal{P} ::= 0 \mid (a,k).\mathcal{P} \mid \mathcal{P} + \mathcal{P} \mid \mathcal{P}\|_S\mathcal{P} \mid \mathcal{P} \setminus L \mid \mathcal{P}' = \mathcal{P}$ where $a \in Act$, $k \in \mathbb{K}$ and $S, L \subseteq Act \setminus \{\tau\}$.*

To avoid too many parenthesis we define among the operations the following decreasing priorities: Hiding, Prefix, Composition and Summation. We also allow for recursion by defining equations, i.e. we use constants as abbreviations by defining $A = P$ which means that A is an abbreviation for P. In contrast to $=$ we use $A \equiv B$ to denote that A and B are syntactically identical. A term in the algebra is denoted as an agent.

The intuitive meaning of the operators is as follows:

- *0* : describes the termination of an agent which afterwards cannot perform any actions.
- *(a,k).A* : action a is performed and afterwards the agent behaves like A, the cost of performing action a equal k.
- $A + B$: the agent behaves either like A or like B.
- $A \|_S B$: A and B proceed concurrently and independently on all actions which are not in S. All actions from S have to be performed as joint actions by both agents.
- $A \setminus L$: actions from the set L are hidden, i.e., they become τ actions which are no longer usable in joint actions with an environment.
- *B=A* : describes that agent B behaves like agent A.

The semantics of an agent $A \in \mathcal{L}$ is a $MLTS$. As usual in process algebras, we cannot distinguish between an agent and a state. An agent and all its derivatives form the state space of a system. We use the notation $A \xrightarrow{a,k} B$ for $T(A, a, B) = k$ if $k \neq \mathbb{O}$. Semantic rules are given in an operational style of the form

$$\frac{premise_1 \ \ldots \ premise_n}{conclusion} \ \langle name \rangle \ (condition)$$

to be read as:
If *condition* is satisfied, the rule $\langle name \rangle$ can be applied and it can be deduced that *conclusion* holds in case of the assumptions $premise_1 \ldots premise_n$ hold.

In all semantic rules we define, conditions are introduced such that transitions with cost $\mathbb{O}$ are not explicitly generated. However, as usual in labeled transitions systems, if no transition between two states exists, then there is implicitly a transition with cost $\mathbb{O}$. Observe that the meaning of $\mathbb{O}$ in all semirings is such that $\mathbb{O} \hat{\cdot} a = \mathbb{O}$ which means that the contribution of a zero element to a vector matrix product is zero. For the prefix and choice operators we have the following operational semantics.

$$\frac{}{(a,k).A \xrightarrow{a,k} A} \; \langle pr \rangle \; (k \neq \mathbb{0}) \qquad \frac{A_j \xrightarrow{a,k} A'}{\sum_{i \in I} A_i \xrightarrow{a,k_\Sigma} A'} \; \langle + \rangle \; \begin{pmatrix} j \in I, \\ k_\Sigma \neq \mathbb{0} \end{pmatrix}$$

where $k_\Sigma = \sum_{i \in I} T(A_i, a, A')$ and I is some index set. Note that the choice operator can yield the situation of multiple arcs with same labels between agents. The above rule ensures that these arcs result as one whose transition function takes a single, well-defined value of k_Σ. For parallel composition, the following four semantic rules are defined.

$$\frac{A \xrightarrow{a,k} A' \quad A' \not\equiv A}{A\|_S B \xrightarrow{a,k} A'\|_S B} \; \langle\|_{S_1}\rangle \; \begin{pmatrix} a \notin S, \\ k \neq \mathbb{0} \end{pmatrix} \qquad \frac{B \xrightarrow{a,k} B' \quad B' \not\equiv B}{A\|_S B \xrightarrow{a,k} A\|_S B'} \; \langle\|_{S_2}\rangle \; \begin{pmatrix} a \notin S, \\ k \neq \mathbb{0} \end{pmatrix}$$

$$\frac{A \xrightarrow{a,k} A \;\; or \;\; B \xrightarrow{a,l} B}{A\|_S B \xrightarrow{a,k \;\hat{+}\; l} A\|_S B} \; \langle\|_{S_3}\rangle \; \begin{pmatrix} a \notin S, \\ k \;\hat{+}\; l \neq \mathbb{0} \end{pmatrix} \qquad \frac{A \xrightarrow{a,k} A' \quad B \xrightarrow{a,l} B'}{A\|_S B \xrightarrow{a,k \;\hat{\cdot}\; l} A'\|_S B'} \; \langle\|_{S_4}\rangle \; \begin{pmatrix} a \in S, \\ k \;\hat{\cdot}\; l \neq \mathbb{0} \end{pmatrix}$$

Rules $\|_{S_1}$, $\|_{S_2}$, and $\|_{S_3}$ describe how both agents proceed independently on actions $a \notin S$, rule $\|_{S_3}$ considers the special case of independent self loops with the same label. The latter is necessary to avoid a reduction of transitions and to retain a well-defined transition function if a parallel composition without synchronisation applies. If only one condition of the premise in $\|_{S_3}$ is satisfied, say $A \xrightarrow{a,k} A'$, then let $l = \mathbb{0}$ for the definition of the conclusion (and vice versa). Actions $a \in S$ are performed as joined actions. The costs of the joint transitions equals the product $\hat{\cdot}$ of the costs of both transitions (rule $\|_{S_4}$) with respect to the semiring. The definition of costs of synchronized transitions is one of the most discussed questions during the development of stochastic process algebras. Several approaches have been proposed in the literature [18] and there are arguments for and against any of these solutions. We do not claim that the use of $\hat{\cdot}$ is always the best way to define the costs. However, it is a natural solution from a mathematical viewpoint and we found reasonable interpretations for this choice in all semirings we considered so far.

The semantics of the hiding operator is defined as

$$\frac{A \xrightarrow{a,k} A'}{A\backslash L \xrightarrow{a,k} A'\backslash L} \; \langle\backslash_1\rangle \; (a \notin L, \; k \neq \mathbb{0}) \; \text{and}$$

$$\frac{A \xrightarrow{a_1,k_1} A' \; ... \; A \xrightarrow{a_n,k_n} A'}{A\backslash L \xrightarrow{\tau,k_\Sigma} A'\backslash L} \; \langle\backslash_2\rangle \; (\{a_1,\ldots,a_n\} \subseteq L \cup \{\tau\}, k_\Sigma \neq \mathbb{0} \,),$$

where $k_\Sigma = \sum_{i=1}^{n} k_i$. Transitions which are not hidden occur as before and hidden transitions between two states appear as a single transition with label τ where costs are accumulated. Note that a transition with label τ can be part of the premises as well.

All operations we have considered so far, allow only the specification of finite behaviors. To define an agent with an infinite behavior one has to define $A = B$ where A occurs in B. E.g., $A = (a, k).A$ is an equation describing an infinite behavior. The semantics of a constant is then given by

$$\frac{A \xrightarrow{a,k} A'}{B \xrightarrow{a,k} A'} \; \langle = \rangle \; (B = A) \, .$$

We sometimes use also variables for processes which can be bound to agents. We use the notation $A\{B/X\}$ when every occurrence of process variable X in A is replaced by agent B. We denote a variable as free if it is not bound to an agent. Following Milner a term of GPA is denoted as an agent if it contains no free variables.

What we have defined in this section is rather typical for a process algebra with an operational semantics. The only difference is that we allow general semirings to define the quantitative values. Concrete applications of this concept are presented in Section 5. The semantics of an agent A is an MLTS where the state space S_A contains A and all its derivatives reachable by applying the semantic rules. The initialization function ini assigns $\mathbb{1}$ to $B \in S_A$ if $B \equiv A$ and $\mathbb{0}$ otherwise. In this way agent A generates an $MLTS$.

Define $\mathcal{D}[A]$ as the set of all successors of agent A, i.e., $\mathcal{D}[A] = \{B \in \mathcal{L} | \exists a \in Act : A \xrightarrow{a,k} B\}$. Similarly $\mathcal{D}_a[A]$ is the set of all successors of A which are reachable by one a labeled transition and $\mathcal{D}^*[A]$ is the transitive and reflexive closure of relation $\mathcal{D}[A]$ and equals S_A.

4   Bisimulation for GPA

Two of the major advantages of process algebras are compositionality and the availability of equivalences which are congruences according to the operations of the algebra. One of the most famous equivalence relations is bisimulation which is well known for untimed systems [24,23] and has been subsequently extended to probabilistic [22] as well as to stochastic systems [7,8,16,19]. In [9] it has been shown that bisimulation can be defined for automata with transition costs where values of costs are elements of a semiring and the behavior of the automaton is defined using vector matrix operations derived from the addition and multiplication of the semiring. Of course, automata with transition costs are identical to finite state MLTS, the semantic model of GPA. We first introduce bisimulation for GPA and prove afterwards the congruence property according to the operations of GPA.

The definition of bisimulation requires some additional notations as a prerequisite. We restrict bisimulations to equivalence relations and assume that the MLTS is of the finite branching type, i.e., the number of transitions with nonzero costs leaving a state is finite. Let $\mathcal{R}$ be an equivalence relation on $\mathcal{L} \times \mathcal{L}$. $CC_{\mathcal{R}}$ is the set of equivalence classes of $\mathcal{R}$, $C_{\mathcal{R}}$ is an equivalence class of $\mathcal{R}$ and $C_{\mathcal{R}}[B]$ is the equivalence class to which agent B belongs.

Definition 4. *An equivalence relation $\mathcal{R}$ on $\mathcal{L} \times \mathcal{L}$ is a bisimulation if and only if for all $(A, B) \in \mathcal{R}$ and all $a \in Act$:*

1. $ini(A) = ini(B)$;
2. *if $A \xrightarrow{a,k} C^1$ for some $C \in \mathcal{L}$, then*

$$\widehat{\sum}_{D \in C_{\mathcal{R}}[C]} T(A, a, D) = \widehat{\sum}_{D \in C_{\mathcal{R}}[C]} T(B, a, D) \ ;$$

---

[1] Observe that $A \xrightarrow{a,k} C$ implies $k \neq \mathbb{0}$ which follows from our semantic rules.

3. if $B \xrightarrow{a,k} C$ for some $C \in \mathcal{L}$, then

$$\sum_{D \in \mathcal{C}_\mathcal{R}[C]} \widehat{T(B,a,D)} = \sum_{D \in \mathcal{C}_\mathcal{R}[C]} \widehat{T(A,a,D)} \ .$$

This definition corresponds to the usual definition of bisimulation for untimed [23], probabilistic [22] or stochastic [8] systems. Note that in the classical bisimulation [23], one considers existence of specific agents C, C' which need to be bisimilar which is the same as to ask for elements $D \in \mathcal{C}_\mathcal{R}[C]$. Like for other bisimulations the following theorem holds.

Theorem 1. Let $\mathcal{R}_1$, $\mathcal{R}_2$ be two bisimulations on $\mathcal{L}$, then $\mathcal{R} = \mathcal{R}_1 \cup \mathcal{R}_2$ is also a bisimulation.

Proof. The proof can be found in [9] and follows the same line of argumentation as the proof for the classical theorem in Boolean systems [23]. □

The theorem implies that the largest bisimulation exists as the union of all bisimulations.

Definition 5. Two agents A and B are bisimilar, denoted as $A \sim B$, if a bisimulation $\mathcal{R}$ exists such that $(A, B) \in \mathcal{R}$.

The term equivalence implies that equivalent agents behave identically in some sense. This is also the case for a bisimilar agent in our general setting. An example is given in the following theorem where we come back to the notion of transition cost described in Section 2 with $c_\omega = \sum_{s \in S} \widehat{\mathbf{p}^\omega(s)}$ as the cost of a transition sequence ω.

Theorem 2. Let A and B be agents generating $MLTS_A$ and $MLTS_B$ with $Act = Act_A = Act_B$. If $A \sim B$, then for each $\omega \in Act^*$, $c_\omega^A = c_\omega^B$ where c_ω^A, c_ω^B are the costs for performing ω in A and B, respectively.

Proof. The proof follows by induction. We first prove that for all $\mathcal{C}_\mathcal{R} \in CC_\mathcal{R}$, $k \in \mathbb{N}$,

$$\sum_{C \in \mathcal{C}_\mathcal{R} \cap \mathcal{D}^*[A]} \widehat{\mathbf{p}^0(C)} = \sum_{D \in \mathcal{C}_\mathcal{R} \cap \mathcal{D}^*[B]} \widehat{\mathbf{p}^0(D)}$$

holds. The relation is true because $\mathbf{p}^0(A) = ini(A) = ini(B) = \mathbf{p}^0(B)$ holds due to $A \sim B$ and Def. 4 and $\mathbf{p}^0(C) = \mathbb{O}$ for all $C \not\equiv A \vee B$ according to the definition of ini for the MLTS semantics of an agent. Now assume that the relation holds for a string ω. We prove the induction to a string $\omega; a$ where $a \in Act$ and $\omega; a$ is the concatenation of ω and a.

$$\sum_{C \in \mathcal{C}_\mathcal{R} \cap \mathcal{D}^*[A]} \widehat{\mathbf{p}^{\omega;a}(C)}$$
$$= \sum_{C'_\mathcal{R} \in CC_\mathcal{R}} \sum_{D \in \mathcal{C}'_\mathcal{R} \cap \mathcal{D}^*[A]} \widehat{\mathbf{p}^\omega(D)} \widehat{\cdot} \left(\sum_{C \in \mathcal{C}_\mathcal{R} \cap \mathcal{D}^*[A]} \widehat{T(D,a,C)} \right)$$
$$= \sum_{C'_\mathcal{R} \in CC_\mathcal{R}} \sum_{D \in \mathcal{C}'_\mathcal{R} \cap \mathcal{D}^*[B]} \widehat{\mathbf{p}^\omega(D)} \widehat{\cdot} \left(\sum_{C \in \mathcal{C}_\mathcal{R} \cap \mathcal{D}^*[B]} \widehat{T(D,a,C)} \right)$$
$$= \sum_{C \in \mathcal{C}_\mathcal{R} \cap \mathcal{D}^*[B]} \widehat{\mathbf{p}^{\omega;a}(C)}$$

The first equation holds due to the cost assignment by a vector matrix multiplication, see (1). The second equation is based on the requirements for a bisimulation (Def. 4). The third equation follows from (1) like the first one. The identity of the vectors implies the identities of the cost functions which can be seen by the following equations.

$$c_\omega^A = \widehat{\sum}_{C_R \in CC_R} \widehat{\sum}_{C \in C_R \cap \mathcal{D}^\bullet[A]} \mathbf{p}^\omega(C) \,\widehat{\cdot}\, \left(\sum_{D \in \mathcal{D}^\bullet[A]} T(C, a, D) \right) =$$

$$\widehat{\sum}_{C_R \in CC_R} \widehat{\sum}_{C \in C_R \cap \mathcal{D}^\bullet[B]} \mathbf{p}^\omega(C) \,\widehat{\cdot}\, \left(\sum_{D \in \mathcal{D}^\bullet[B]} T(C, a, D) \right) = c_\omega^B$$

$\square$

The theorem introduces some form of trace equivalence for GPA which is observed by bisimilar agents.

An equivalence relation is especially useful, if it is a congruence according to the operations of the algebra. In this case, equivalence assures that equivalent agents behave "equivalently" in all possible environments and equivalent agents can be substituted which allows the interleaving of composition and aggregation [11,8]. The following theorems 3, 4 state that bisimulation is indeed a congruence according to the operations of GPA.

Theorem 3. *Let $A, B, C \in \mathcal{L}$ with $A \sim B$, then the following relations hold*

1. *$(a, k).A \sim (a, k).B$ for all $a \in Act$ and $k \in \mathbb{K} \setminus \{\mathbb{O}\}$,*
2. *$A + C \sim B + C$,*
3. *$A \setminus L \sim B \setminus L$ for all $L \subseteq Act \setminus \{\tau\}$, and*
4. *$A\|_S C \sim B\|_S C$ for all $S \subseteq Act \setminus \{\tau\}$.*

The proofs follow from the proofs for untimed process algebras [23] or stochastic process algebras [8]. Note that we have a strong similarity to untimed process algebras if one focuses only on labels $a \in Act$, such that the crucial point for bisimulation is in the treatment of elements of the semiring. A key observation is that the semantics of Prefix, Choice and Hiding preserve existence of transitions and that if multiple transitions between two states are merged into a single one, elements of the semiring are added by $\widehat{+}$. This matches requirements for the bisimulation which adds elements of the semiring by $\widehat{+}$ as well. Due to associativity and commutativity of $\widehat{+}$ in Def. 1 the order in which elements are added does not matter. We omit proofs of prefix, choice and hiding for Theorem 3 since these are lengthy and do not give further insight. We focus on the proof of composition, since this one is the most interesting, especially rule $\|_{S4}$, where elements of the semiring are multiplied. Observe that the distributive laws of the semiring allow to conclude that for finite index sets I, J, and K and a_i, b_j, c_k elements of the semiring for all $i \in I$, $j \in J$, $k \in K$ holds

$$\widehat{\sum}_{i \in I, k \in K} a_i \,\widehat{\cdot}\, c_k = \left(\widehat{\sum}_{i \in I} a_i \right) \widehat{\cdot} \left(\widehat{\sum}_{k \in K} c_k \right) = \left(\widehat{\sum}_{j \in J} b_j \right) \widehat{\cdot} \left(\widehat{\sum}_{k \in K} c_k \right)$$
$$= \widehat{\sum}_{j \in J, k \in K} b_j \,\widehat{\cdot}\, c_k$$

if $\widehat{\sum}_{i \in I} a_i = \widehat{\sum}_{j \in J} b_j$. Having this in mind, the argumentation on equations for proving rule $\|_{S4}$ are easier to follow. The necessity of associativity and distributivity of $\hat{\cdot}$ over $\hat{+}$ for existence of congruences is well known, e.g. [18]. The proof follows the same line of argumentation as Milner's proof of strong bisimulation in [23].

Proof. of composition in Theorem 3

Let $\mathcal{R}$ be an equivalence relation such that $(A\|_S C, B\|_S C) \in \mathcal{R}$ if and only if $A \sim B$. We have to prove that $\mathcal{R}$ is a bisimulation. Since $ini(A\|_S C) = ini(B\|_S C) = \mathbb{1}$ and $ini(D) = \mathbb{0}$ for $D \not\equiv (A\|_S C) \vee (B\|_S C)$ by definition of the MLTS generated by $A\|_S C$, resp. $B\|_S C$, the first condition of a bisimulation is satisfied.

Suppose $(A\|_S C, B\|_S C) \in \mathcal{R}$, let $A\|_S C \xrightarrow{a,k} E$ then there are four cases according to the four semantic rules of composition; cases 1-3 consider $a \notin S$:

Case 1: $A \xrightarrow{a,k} A'$, $A \not\equiv A'$ and $E \equiv A'\|_S C$

We use the notation $D \in C_\sim[A']$ as in Def. 4 with $\sim$ as symbol $\mathcal{R}$ there. Since $A \sim B$ we have $\mathbb{0} \neq \widehat{\sum}_{D \in C_\sim[A']} T(A, a, D) = \widehat{\sum}_{D \in C_\sim[A']} T(B, a, D)$ so we have $D \in C_\sim[A']$ where $B\|_S C \xrightarrow{a,l} D\|_S C$ and $(A'\|_S C, D\|_S C) \in \mathcal{R}$.

Case 2: $C \xrightarrow{a,k} C'$, $C \not\equiv C'$ and $E \equiv A\|_S C'$

Then also $B\|_S C \xrightarrow{a,k} B\|_S C'$ and $(A\|_S C', B\|_S C') \in \mathcal{R}$.

Case 3: $A \xrightarrow{a,l} A$ or $C \xrightarrow{a,m} C$ and $E \equiv A\|_S C$, $k = l \hat{+} m$.

If only one of the two conditions holds, then we have the same argumentation as for case 1 or 2, respectively. If both agents contain a self loop with label a, then from $A \sim B$ we have $\mathbb{0} \neq \widehat{\sum}_{D \in C_\sim[A]} T(A, a, D) = \widehat{\sum}_{D \in C_\sim[A]} T(B, a, D)$. So for $D \in C_\sim[A]$ we have $B \xrightarrow{a,n} D$, but since $A \in C_\sim[A]$ and $A \sim B$, also $B \in C_\sim[A]$, such that $(A\|_S C, B\|_S C) \in \mathcal{R}$ and

$$\widehat{\sum}_{D \in C_\sim[A]} T((A\|_S C), a, (D\|_S C)) \hat{+} \widehat{\sum}_{F \in C_\sim[C]} T((A\|_S C), a, (A\|_S F)) =$$

$$\widehat{\sum}_{D \in C_\sim[B]} T((B\|_S C), a, (D\|_S C)) \hat{+} \widehat{\sum}_{F \in C_\sim[C]} T((B\|_S C), a, (B\|_S F))$$

due to the associative and commutative law for $\hat{+}$ in the semiring.

Case 4: $a \in S$, $A \xrightarrow{a,l} A'$, $C \xrightarrow{a,m} C'$ and $E \equiv A'\|_S C'$, $k = l \hat{\cdot} m$.

By $A \sim B$ we have $B \xrightarrow{a,n} B'$ with $B' \in C_\sim[A']$ so clearly $B\|_S C \xrightarrow{a,o} B'\|_S C'$ and $(A'\|_S C', B'\|_S C') \in \mathcal{R}$. It remains to show equality of transition costs in the semiring. For each $C_\mathcal{R} \in CC_\mathcal{R}$ we have:

$$\widehat{\sum}_{(D\|_S F) \in C_\mathcal{R}} T((A\|_S C), a, (D\|_S F)) = T(C, a, F) \hat{\cdot} \widehat{\sum}_{C_\sim \in CC_\sim} \widehat{\sum}_{D \in C_\sim} T(A, a, D)$$
$$= T(C, a, F) \hat{\cdot} \widehat{\sum}_{C_\sim \in CC_\sim} \widehat{\sum}_{D \in C_\sim} T(B, a, D) = \widehat{\sum}_{(D\|_S F) \in C_\mathcal{R}} T((B\|_S C), a, (D\|_S F))$$

due to distributive laws for $\hat{+}$ and $\hat{\cdot}$ in the semiring. This shows that the identity of transition costs holds. By a symmetric argument we complete the proof. □

It remains to show that bisimulation is a congruence for recursive behaviors. We first define under which conditions agents including variables are bisimilar.

Definition 6. *Let $A, B \in \mathcal{L}$ and let A, B include variables X_i ($i \in I$ for index set I) at most. $A \sim B$ if and only if for all agents C_i ($i \in I$)*

$$A\{C_i/X_i\} \sim B\{C_i/X_i\} .$$

Theorem 4. *Let $C, D \in \mathcal{L}$, $C \sim D$ and let C and D both contain variable X. Let $A = C\{A/X\}$ and $B = \{B/X\}$, then $A \sim B$.*

Proof. Since recursion by defining equations does not involve elements of the semiring, the proof follows from the proof for the Boolean model, e.g. as given in [23]. □

By means of the above theorems equivalence transformations can be defined at the syntactical level similar to the rules in process algebras like CCS [23].

5   Examples and Realizations

It is straightforward to define different process algebras using the general concept which has been presented in the previous sections. We give three application examples in the following subsections.

5.1   A Probabilistic Process Algebra

Different realizations of probabilistic processes exist in the literature [25,22]. On particular problem with probabilistic processes occurs when composition is introduced because after composition the sums of probabilities of outgoing transitions for an agent might be smaller or larger than 1. Some calculi introduce rescaling of probabilities in such a case which, however, may yield some problems with an appropriate definition of bisimulation as shown in [12]. We avoid this problem by imposing sufficient restrictions on our definition of a probabilistic process algebra for the semiring $(\mathbb{R}_+, +, \cdot, 0, 1)$. For each agent A and each $a \in Act$ the following condition holds.

$$\sum_{B \in \mathcal{D}_a[A]} T(A, a, B) \in \{0, 1\} \tag{2}$$

Thus an agent can either perform action a or it cannot perform this action. If an action is possible, then the successor agent might be chosen probabilistically. In the notation of [25] this describes a reactive system. We assume in the sequel that Act does not contain label τ and all labels can be used for synchronization. Note that hiding makes it difficult to retain property (2).

Corollary 1. *Let A and B be two agents observing (2) and let Act_A and Act_B not contain label τ, then the following agents also observe (2):*

- $(a, 1).A + (b, 1).B$ with $a \neq b$,
- $(a, p).A + (a, 1 - p).B$ for $0 < p < 1$,
- $A + B$ if $Act_A \cap Act_B = \emptyset$,
- $A\|_S B$ if $(Act_A \cap Act_B) \setminus S = \emptyset$

and any combination of the above agents.

Reactive systems are usually driven by some scheduler a component which resolves the nondeterminism by choosing actions to perform. An agent AS is a scheduler if for all $A \in \mathcal{D}^*[AS]$ the following condition holds.

$$\sum_{a \in Act \setminus \{\tau\}} \sum_{B \in \mathcal{D}[A]} T(A, a, B) \in \{0, 1\} \tag{3}$$

An agent $A \in \mathcal{D}^*[AS]$ is terminating if $\mathcal{D}[A] = \emptyset$. Each composition $AS\|_{Act}B$ where AS observes (3) and B observes (2) is stochastic process where the sum of outgoing probabilities is 1.0 or 0.0 for each agent. The resulting process reaches a terminating state if the scheduler reaches a terminating state and it reaches a deadlock state if it reaches a state without successors but the scheduler is not in a terminating state. Alternatively, one might be interested in an infinite behavior which means that the system never reaches a state without successors.

As a simple example we consider the well known dining philosophers problem where philosophers pick up one fork after the other, but a philosopher may first pick up the left or the right fork depending on their availability. We consider a system with N philosophers and forks where philosopher n and fork n are described by the following terms in GPA.

$$PH_n^1 = (g_n^l, 1).(g_n^r, 1).PH_n^2 + (g_n^r, 1).(g_n^l, 1).PH_n^2,$$

$$PH_n^2 = (p_n^l, 1).(p_n^r, 1).PH_n^1 + (p_n^r, 1).(p_n^l, 1).PH_n^1,$$

$$F_n = (g_n^r, 1).(p_n^r, 1).F_n + (g_{n+1}^l, 1).(p_{n+1}^l, 1).F_n$$

where $n + 1$ in g_{n+1} and p_{n+1} is performed modulo N to have a cyclic system. The overall system results from composition of philosophers PH_n^1, forks F_n and a scheduler AC with $S = \cup_{i=1}^n \{g_i^l, g_i^r, p_i^l, p_i^r\}$. Two possible schedulers which allow an infinite behavior of the system are the following two.

$$AC_1 = (g_1^r, 1/N).(g_1^l, 1).(p_1^l, 1).(p_1^r, 1).AC_1 + \ldots$$
$$+ (g_N^r, 1/N).(g_N^l, 1).(p_N^l, 1).(p_N^r, 1).AC_1$$

$$AC_2 = (g_1^l, 1).(g_1^r, 1).(p_1^l, 1).(p_1^r, 1). \ldots .(p_N^l, 1).(p_N^r, 1).AC_2$$

Of course, other schedulers allowing more parallelism can be defined as well.

5.2   Max/Plus Process Algebra

As a second example we consider GPA over the semiring $(\mathbb{R} \cup \{-\infty\}, \max, +, -\infty, 0)$. To the best of our knowledge no process algebra has been proposed for this semiring yet. However, this model is well suited for the formulation of deterministic scheduling problems or several other optimization problems.

Deterministic scheduling problems are often formulated using marked graphs, a specific class of Petri nets without choices. As an alternative formalism one may as well describe these systems using GPA with the advantage of compositionality and the availability of a well defined equivalence. For the application of max/+ algebra to the analysis of timed marked graphs we refer to [3].

Max/+ algebra can only be applied for the analysis of Petri nets without choices. This restriction is, of course, taken over to GPA. Thus, an agent for the analysis of a scheduling problem must not include choices which means that the + operation cannot be used for the specification of agents.

Since the formulation of models in GPA over the max/+ semiring is straightforward, we consider a simple example. The example consists of two sources which generate parts of raw material to be assembled by a machine. The two sources are described by agent A and B and generate their material after constant times t_A and t_B.

$$A = (\tau, t_A).A_1 \quad A_1 = (a, t_p).A \quad B = (\tau, t_B).B_1 \quad B_1 = (a, t_p).B$$

The parts are assembled by a machine described by component C. The machine picks up the parts if both are available by a transition with label a where picking takes time t_p for each part. It subsequently assembles both parts and offers the assembled part to some environment via a transition with label b. We assume that the machine has no intermediate buffer such that the assembled part has to be first delivered before processing of a new part can start.

$$C = (a, 0).C_1 \quad C_1 = (\tau, t_C).C_2 \quad C_2 = (b, 0).C$$

The whole system is composed of the three components. Label a is hidden because the environment interacts with the system only via b.

$$Sys = ((A\|_a B)\|_a C) \setminus \{a\}$$

The set $\mathcal{D}^*[Sys]$ contains the following 12 states which are denoted by the state of the components.

1) (A, B, C) 2) (A, B, C_1) 3) (A, B, C_2) 4) (A, B_1, C)
5) (A, B_1, C_1) 6) (A, B_1, C_2) 7) (A_1, B, C) 8) (A_1, B, C_1)
9) (A_1, B, C_2) 10) (A_1, B_1, C) 11) (A_1, B_1, C_1) 12) (A_1, B_1, C_2)

The MLTS of the system is shown in Fig. 1. In the picture, transition labels are not shown instead τ-labeled transitions are denoted by solid arcs, whereas b-labeled transitions are described by dashed arcs. Transition costs are written near the arcs. Observe that $\mathbb{O} = -\infty$ in this semiring such that all arcs which are not shown have weight $-\infty$. If $t_A = t_B$, then the relation $\mathcal{R} = \{(1), (2), (3), (4, 7), (5, 8), (6, 9), (10), (11), (12)\}$ is a bisimulation for the system and a smaller equivalent system can be generated which contains 9 instead of 12 states. Equivalent agents are surrounded by dotted boxes in the figure. The system can be analyzed according to the completion time of parts by computing vector matrix products in the general form described above.

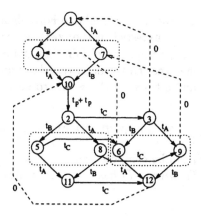

Fig. 1. MLTS of the scheduling example.

5.3 Min/Plus Process Algebra

Max/+ algebra is often used for systems with synchronization, where weights describe delays such that at a synchronization point one agent has to wait for another to do the synchronization. In min/+ algebra, $\widehat{+}$ is becomes minimum. This is useful to formulate optimization problems like shortest path problems or Markov decision problems. Of course, Min/+ and Max/+ are very similar since one can substitute the other by simply using negative values.

As an example for min/+ systems we consider a model for the computation of minimum traveling costs. The scenario is simple: A traveler wants to go from a source place to a destination place. Between source and destination several cities are located and the traveler can choose different ways and different means of transport. We use label b for buses, t for trains and c for a cab and consider as an example a route with 6 cities. City 1 is the source, city 7 is the destination which is denoted as agent 0 in our specification. The following agent defines the connections between cities.

$$C_0 = (b, 2).C_2 + (b, 2).C_4 + (c, 6).C_5 \quad C_1 = (b, 1).C_0 + (t, 2).C_5$$
$$C_2 = (t, 2).C_1 + (c, 3).C_3 \qquad\qquad C_3 = (c, 5).0 + (t, 1).C_4 + (c, 4).C_5$$
$$C_4 = (t, 4).0 + (c, 3).C_5 \qquad\qquad C_5 = (b, 2).0 + (b, 2).C_3 + (t, 1).C_4$$

A traveler can now be defined to synchronize with agent C_0. Consider first a traveler who can use arbitrary means of transport.

$$T_1 = (b, 0).T_1 + (c, 0).T_1 + (t, 0).T_1$$

The agent $Sys_1 = (C_0\|_{b,c,t}T_1) \setminus \{b, c, t\}$ can be used to determine the minimum traveling costs. The MLTS for this system contains 7 states. Initially $C_0\|_{b,c,t}T$ receives cost $\mathbb{1} = 0$ and all remaining states receive costs $\mathbb{0} = \infty$. Costs can be computed by computing vectors $\mathbf{p}^k$; if $\mathbf{p}^k = \mathbf{p}^{k+1}$, then $\mathbf{p}^k(A)$ contains the minimal costs of reaching agent $A \in \mathcal{D}^*[Sys_1]$ from the initial state. The used algorithm is, of course, the well known Bellman Ford algorithm for shortest path problems [6]. Since we are interested in the costs of reaching the destination, $\mathbf{p}^k(0)$ is of interest. In this example we obtain minimal costs of 6.

We may as well define other travelers. A traveler who is not allowed to take a taxi is defined as T_2 and a traveler who wants to move at most once by train or bus is defined as T_3.

$$T_2 = (b,0).T_1 + (t,0).T_2 \quad T_3 = (c,0).T_3 + (b,0).T_{31} + (t,0).T_{31} \quad T_{31} = (c,0).T_{31}$$

Both agents, T_2 and T_3 can be composed with C_0 like T_1. For T_2 we obtain the same minimum as for T_1, wheras T_3 has costs of 7 which is the additional price to use a cab.

6   Conclusions

In this paper we proposed a new and generic approach to define process algebras with quantified transitions. It has been shown that using an arbitrary semiring structure it is possible to define a process algebra with transition costs described by the elements of the semiring. Different behaviors of an agent are expressed by the operations of the semiring. The general process algebra GPA has been introduced based on the presented concepts. A very interesting aspect of this general approach is that a bisimulation equivalence can be defined based only on the requirements of the semiring and that the equivalence is a congruence according to the operations of the algebra.

It has been shown that by choosing concrete semirings we obtain process algebras very similar to existing process algebras. In this way it is easy to define untimed, probabilistic or stochastic algebras as it has already been done. We do not elaborate on the derivation of a stochastic algebra in our framework. Stochastic algebras mainly differ in the selection of rates in case of synchronisation [18]. Since composition is based on $\hat{\cdot}$ in GPA, selection of semiring $(\mathbb{R}_+, +, \cdot, 0, 1)$ yields a stochastic algebra in the flavor of MTIPP [16]. Rather than deriving existing approaches, we use other semirings to achieve new process algebras including an appropriate notation of bisimulation. In this paper we present a max/+ and a min/+ process algebra which are useful for the formulation of various optimization problems. The semirings max/+ and min/+ have been considered here only for constant values. The resulting model is commonly used for the analysis of deterministic problems [3,10]. Additionally, max/+ algebra is applied for the analysis of stochastic systems [2,10] by considering non-decreasing functions instead of constants as elements of the semiring. It is obvious that most of the steps proposed in this paper can be taken over to this more general model. However, some extensions have to be introduced and are considered in future research.

References

1. R. Agrawal, F. Baccelli, and R. Rajan. An algebra for queueing networks with time varying service. Research Report 3435, INRIA, 1998.

2. F. Baccelli, G. Cohen, G. Olsder, and J. Quadrat. *Synchronization and Linearity.* John Wiley and Sons, 1992.
3. F. Baccelli, B. Gaujal, and D. Simon. Analysis of preemptive periodic real time systems using the (max,plus) algebra. Research Report 3778, INRIA, 1999.
4. F. Baccelli and D. Hong. TCP is (max/+) linear. In *Proc. SIGCOM 2000.* ACM, 2000.
5. M. Bernado and R. Gorrieri. A tuturial of EMPA: A theory of concurrent processes with nondeterminism, priorities, probabilities and time. *Theoretical Computer Science,* 202:1–54, 1998.
6. D. P. Bertsekas and J. N. Tsitsiklis. *Parallel and distributed computation.* Prentice Hall, 1989.
7. P. Buchholz. *Die strukturierte Analyse Markovscher Modelle (in German).* IFB 282. Springer, 1991.
8. P. Buchholz. Markovian process algebra: composition and equivalence. In [17].
9. P. Buchholz. Bisimulation for automata with transition costs. *submitted,* 2000.
10. C. S. Chang. *Performance guarantess in communication networks.* Springer, 1999.
11. R. Cleaveland, J. Parrow, and B. Steffen. The concurrency workbench: a semantics based tool for the verification of concurrent systems. *ACM Transactions on Programming Languages and Systems,* 15(1):36–72, 1993.
12. P. R. D'Argenio, H. Hermanns, and J. P. Katoen. On generative parallel composition. In *Proc. ProbMiv 98, Electronic Notes on Theor. Computer Sc.,* 21, 1999.
13. D. Gross and D. Miller. The randomization technique as a modeling tool and solution procedure for transient Markov processes. *Operations Research,* 32(2):926–944, 1984.
14. H. Hansson. Time and probability for the formal design of distributed systems. Phd thesis, University of Uppsala, 1991.
15. H. Hermanns, U. Herzog, and V. Mertsiotakis. Stochastic process algebras – between LOTOS and Markov chains. *Computer Networks and ISDN Systems,* 30(9/10):901–924, 1998.
16. H. Hermanns and M. Rettelbach. Syntax, semantics, equivalences, and axioms for MTIPP. In [17]
17. U. Herzog and M. Rettelbach, eds. *Proc. 2nd Work. Process Algebras and Performance Modelling* Arbeitsberichte IMMD, Univ. Erlangen, Germany, no. 27, 1994.
18. J. Hillston. The nature of synchronisation. In [17].
19. J. Hillston. A compositional approach for performance modelling. Phd thesis, University of Edinburgh, Dep. of Comp. Sc., 1994.
20. J. Hillston. Compositional Markovian modelling using a process algebra. In W. J. Stewart, editor, *Computations with Markov Chains,* pages 177–196. Kluwer, 1995.
21. C. Hoare. *Communicating sequential processes.* Prentice Hall, 1985.
22. K. Larsen and A. Skou. Bisimulation through probabilistic testing. *Information and Computation,* 94:1–28, 1991.
23. R. Milner. *Communication and concurrency.* Prentice Hall, 1989.
24. D. Park. Concurrency and automata on infinite sequences. In *Proc. 5th GI Conference on Theoretical Computer Science,* pages 167–183. Springer LNCS 104, 1981.
25. R. van Glabbek, S. Smolka, B. Steffen, and C. Tofts. Reactive, generative and stratified models for probabilistic processes. In *Proc. LICS'90,* 1990.

Implementing a Stochastic Process Algebra within the Möbius Modeling Framework*

Graham Clark and William H. Sanders

Dept. of Electrical and Computer Engineering and Coordinated Science Laboratory
University of Illinois at Urbana-Champaign
1308 W. Main St., Urbana, IL, USA

Abstract. Many formalisms and solution methods exist for performance and dependability modeling. However, different formalisms have different advantages and strengths, and no one formalism is universally used. The Möbius tool was built to provide multi-formalism multi-solution modeling, and allows the modeler to develop models in any supported formalism. A formalism can be implemented in Möbius if a mapping can be provided to the Möbius Abstract Functional Interface, which includes a notion of state and a notion of how state changes over time. We describe a way to map PEPA, a stochastic process algebra, to the abstract functional interface. This gives Möbius users the opportunity to make use of stochastic process algebra models in their performance and dependability models.

1 Introduction

Many performance and dependability modeling formalisms and model solution methods have been developed. The most suitable formalism for a particular model depends on many factors, including the experience of the modeler, the results required, and the resources available to solve the model. In addition, large modeling projects may be split among several teams, each with different modeling backgrounds. For these reasons, it is desirable to provide techniques and tools for constructing heterogeneous models that may be composed of sub-models of different formalisms, each seamlessly interacting with each other and with model solvers. The Möbius project aims to provide such techniques and tools.

The theory of Möbius is designed to support heterogeneous modeling by the use of an *abstract functional interface* (AFI [1,2]), a specification that any candidate modeling formalism must implement. By doing so, the formalism ensures that its models may share state and synchronize with other models, support the definition of performance variables, and be solved using one of several

---

* This material is based upon work supported in part by the National Science Foundation under Grant No. 9975019 and by the Motorola Center for High-Availability System Validation at the University of Illinois (under the umbrella of the Motorola Communications Center).

L. de Alfaro and S. Gilmore (Eds.): PAPM-PROBMIV 2001, LNCS 2165, pp. 200–215, 2001.

methods. The first formalism supported by Möbius was stochastic activity networks (SANs [3]). This paper gives details of the incorporation of a new modeling formalism, PEPA [4], which is a stochastic process algebra (SPA). Our work means that PEPA models may now be specified, composed with other submodels, and solved within the Möbius tool.

Section 2 provides some background on the Möbius framework and tool, and a review of the required SPA concepts. Section 3 discusses equivalence-sharing in Möbius, and some basic notions of state for a PEPA process. Section 4 presents an extension to PEPA that makes use of process parameters, and gives details of how this extension can be employed in providing a useful and intuitive mapping to the Möbius AFI. Section 6 presents an example of modeling with PEPA using the Möbius tool, and finally Section 7 discusses ways in which the Möbius framework could be extended in the future, and concludes the paper.

2   Background

In this section, we first describe the Möbius framework and its implementation as the Möbius tool. Following that, we provide a reminder of the relevant features of our stochastic process algebra, PEPA.

2.1   The Möbius Framework

The Möbius framework [5,6] provides an abstract set of requirements for building a particular modeling formalism. It is based upon a theory that is motivated by many existing modeling formalisms, and seeks to capture the essential components of models built using these formalisms. Any model that is built according to the Möbius framework must present a specified interface to other models and solvers within the framework. The implementation of this interface is known as the *AFI*. The AFI includes:

- a set S of *state variables*, and
- a set A of *actions*.

As described in [6], a state variable consists of a type, a value, and a distribution over initial values. Its value is typically used to represent the state of a component or subcomponent, such as the number of customers at a service center in a queuing network. The Möbius framework specifies a rich and structured set of variable types T. In particular, the integers and subsets of T are all state variable types. In theory this set is of infinite size; in practice we assume that it is arbitrarily large, but finite. Of course, this is a reasonable assumption when it comes to implementing the Möbius tool. The type of a state variable is given by a function $type : S \rightarrow T$; the value of a state variable is given by a function $val : S \rightarrow V$ where S is the set of all state variables, and $V \in T$. Two state variables s_1 and s_2 are *compatible* if and only if $type(s_1) = type(s_2)$, and are equal if and only if they are compatible, and $val(s_1) = val(s_2)$.

An action consists of a set of action functions (and in general, some state of its own, used when building the stochastic process of the model). An action is responsible for changing the values of state variables. Just as state variables provide an abstraction of model state, actions are intended to be an abstraction of the state-changing methods of existing modeling formalisms. For a given action $a \in A$, the action function $Enabled_a : V \to bool$ determines whether or not a is "active" in the current state and capable of changing the state at some point in the future if uninterrupted. $Delay_a : V \to (\mathbb{R} \to [0,1])$ associates a probability distribution function (PDF) with an action in the current state; this PDF describes the time from the point of enabling until completion. Given a state, $Complete_a : V \to V$ specifies the state that will result from the completion of the action. Because every constructed Möbius model is intended for performance and dependability evaluation, non-probabilistic non-determinism is not directly supported in the AFI. There are more subtle action functions that capture the effect that action interruptions have on delay distributions; for more details, see [6].

2.2   The Möbius Tool [2]

Models are built hierarchically with Möbius. The modeler begins by specifying a set of one or more *atomic* models. For example, one of these atomic models may be a SAN, just as would have been supplied to the Möbius tool's predecessor, *UltraSAN* [7]. A *composed model* is built from a set of atomic (or indeed composed) models. The Möbius tool currently features two composed model formalisms, both based on the notion of *equivalence sharing*. In both formalisms, submodels are linked together such that compatible state variables may be identified. The AFI ensures that the formalism need not know the implementation details of its submodels' state. We refer to a set of composed submodels as *partner models*. A benefit of using these composed model formalisms is that the theory of *reduced base model construction* [8] is employed to construct a lumped stochastic process, which often results in significantly smaller state spaces. An atomic or composed model is then combined with a *performability variable* defined on the model to generate a *solvable model*. The performability variable is a description of the particular measure that the modeler wishes to calculate. A solvable model may be parameterized on a set of *global variables* to produce a *study* which is composed of a set of *experiments*. Current work is focusing on inferring statistics from a constrained set of experiments using a design of experiments approach [9]. Finally, the modeler must choose a technique for solving the collection of experiments, and the particular solver to use. The Möbius tool provides a number of analytical solvers that use a variety of linear algebra techniques for solving for steady-state and transient measures. Alternatively it is possible to employ an efficient discrete event simulator, which provides confidence intervals for the required performance measures.

2.3   Stochastic Process Algebras and PEPA

In recent years, interest has grown in the use of process-algebra-based methodologies for performance modeling and evaluation. PEPA [4] is a well-known stochastic process algebra. Process algebra notations are based on formal languages, and the PEPA language provides a small set of combinators, presented below:

$$S ::= (\alpha, r).S \mid S + S \mid A_S$$
$$P ::= P \underset{L}{\bowtie} P \mid P/L \mid A \mid S$$

Any process described by S is termed a *sequential component*. A process P consists of a *model configuration* of sequential or model components. A PEPA model consists of a set of definitions and a *system equation*, a distinguished PEPA term that can be interpreted as the starting state. The language is deliberately parsimonious as this helps keep the theory manageable and reduce the proof burden.

For a detailed description of PEPA's combinators, see [4]. *Prefix* is the most fundamental combinator; a process $(\alpha, r).P$ may perform activity (α, r), which has *action type* α and is exponentially distributed with mean $1/r$, and then evolve into process P. We use a to denote an arbitrary activity. Process $P + Q$ expresses a competitive *choice* between P and Q in which the enabled activities of P and Q compete, and a race condition distinguishes the component into which the process evolves. The *cooperation* $P \underset{L}{\bowtie} Q$ is a process that expresses the parallel and synchronizing execution of both P and Q. Both components proceed independently on activities whose types are not in the set L. However, those activities with types captured by L require the concurrent participation of both subprocesses, and this results in an activity with a rate that reflects the rate of the slower participant. Finally, P/L is the process that *hides* activities with types in L. These become *silent* activities with type τ. This combinator is used to express abstraction.

Processes may be recursively defined as the least solution to a set of equations where each is of the form $A \stackrel{\text{def}}{=} P$. A classical process algebra combinator missing from PEPA is the nullary combinator 0, representing the *deadlocked* process. To date, the focus with PEPA modeling has been on steady-state measures, for which 0 has little application. 0 can still be represented in PEPA (for example, if $P \stackrel{\text{def}}{=} (\alpha, r).P$, $Q \stackrel{\text{def}}{=} (\beta, s).Q$, then $P \underset{\{\alpha,\beta\}}{\bowtie} Q$ is deadlocked). We make use of a deadlocked process later in the paper. The operational semantics of PEPA infer the *transitions* of a compound process from the transitions of its subcomponents. If $P \xrightarrow{(\alpha, r)} P'$ then P' is called an $((\alpha, r)\text{-})derivative$ of P. If $P \xrightarrow{(\alpha, r)}$, then there exists some P' such that P' is an (α, r)-derivative of P. The *derivative set* of a PEPA process P is denoted by $ds(P)$, and is the smallest set of components that is closed under the transitive closure of the transition relation. This captures all "reachable states" from P. Both prefix and choice are termed *dynamic* combinators, meaning that the combinators do not persist (in general) over transitions. In contrast, cooperation and hiding do persist over transitions, and are termed *static.* From the transition system, a Markov chain can be produced by essentially discarding activity labels on arcs, providing a performance model. In order to perform a steady-state analysis with a finite state space, it is required that the underlying Markov chain be irreducible and ergodic. A necessary (but not sufficient) condition for this is that the PEPA process be *cyclic.* A *cyclic* PEPA process is a process with a structure such that no static combinator is within the scope of a dynamic combinator. Since static combinators are never "destroyed" by activity transitions, this syntactic condition ensures that the structure of the process does not grow unboundedly over time. For more details on PEPA, including its well-developed equational theory, see [4].

3   Equivalence Sharing

In the Möbius framework, models M_1 to M_n exhibit an *equivalence sharing* relationship if there exists a state variable s_i from each model such that for $1 \leq i \leq j \leq n$, s_i and s_j are compatible, and at all times, $val(s_i) = val(s_j)$. Möbius uses equivalence sharing relationships in the construction of Replicate/Join and Graph composed models. The Möbius tool first uses state identification and modification methods provided in the implementation of the AFI to link together the appropriate portions of submodel state. Any component of Möbius that requests information about the state of the composed model must also make use of the AFI, ensuring that the correct data is returned. In this way, Möbius presents a uniform view of model state, but allows for internal efficiencies in storing composed model state. Equivalence sharing allows for what can be viewed as a form of "two-way communication" between models. By altering the value of the shared portion of model state, one submodel can influence the behavior of its partners, and similarly, can have its behavior influenced by its partners.

3.1   Representations of State

PEPA's operational semantics provide a translation from a collection of algebraic expressions into a graph model, which leads to a continuous-time Markov chain. Each state of the Markov chain is associated with a particular process expression (or an equivalence class of process expressions). Therefore, the PEPA process can be viewed as evolving over time according to transition rules, with the current state represented by a particular term of the process algebra. As mentioned in Section 2.3, we restrict ourselves to considering cyclic PEPA processes only, in order to prevent the structure of the PEPA terms from growing without bound over time. Since we consider only cyclic PEPA processes, we can view a PEPA process as having a tree structure such that no "static" nodes lie below a "dynamic" node. Since the static structure is invariant over activity transitions, we consider everything at the level of, and below, a dynamic combinator to be an evolving subcomponent. A simple tree representation is presented in Figure 1, in which the dotted rectangles highlight the subcomponents.

$$Sys \stackrel{\text{def}}{=} (((\alpha, r).P' \underset{\{\alpha\}}{\bowtie} (\alpha, s).Q')/\{\alpha\}) \underset{L}{\bowtie} (R + S)$$

Taken together, the individual states of these subcomponents, along with the invariant static structure of the process, are enough to characterize the state of the model as a whole.

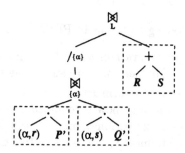

Fig. 1. Highlighting submodel state for *Sys*

A simple vector notation for *Sys* is $\langle (\alpha, r).P', (\alpha, s).Q', R + S \rangle_{Sys}$. The states of each sequential component may be enumerated, leading to a simple first mapping to a set of state variables for the AFI. Concretely, for $\langle R_1, \ldots, R_n \rangle_P$ and $1 \le i \le n$, state variable s_i is such that $type(s_i) = \mathbb{N}$ and $val(s_i) = enum(R_i)$. However, after considering the use of these state variables in an equivalence sharing relationship, we reject this as a workable mapping for two reasons:

- There is no obviously meaningful way to interpret the natural number enumeration of a sequential component, and there seems to be no compelling way in which a partner model could use this data.
- If a partner model changed the value of one or more state variables, this would effect a "jump" in the structure of the PEPA model. This is too uncontrolled.

Instead, our technique will rely on extending PEPA with *process parameters* and providing these as state variables. We next present PEPA$_k$, our extension to PEPA.

4   Extending PEPA with Process Parameters

In this section, we present PEPA$_k$, an extension to PEPA that makes use of process parameters. Extending process algebras with parameters is certainly not a new idea; for example, see [10,11,12,13]. What is novel is our use of this extension to implement equivalence sharing between models. We present the syntax of PEPA$_k$ next.

Definition 1 (Syntax of PEPA$_k$) *Let A range over a set C of* process constants, A_S *over a set $C_S \subseteq C$ of* sequential process constants, *and x over a set*
X *of* process parameters. *Let e represent the syntax of arithmetic expressions
over X, and b represent the syntax of predicates over X. Then the syntax of
PEPA$_k$ is given by:*

$$S ::= (\alpha, r).S \mid (\alpha!e, r).S \mid (\alpha?x, r).S \mid S + S \mid \text{if } b \text{ then } S \mid A_S \mid A_S[\underline{e}]$$
$$P ::= P \underset{L}{\bowtie} P \mid P/L \mid A \mid A[\underline{e}] \mid S$$

PEPA$_k$ provides the following additions to PEPA:

Formal Parameters: process variables now have an arity and may be instantiated with parameters. Furthermore, processes may be specified through the
definition of equations of the form $P[x_1, \ldots, x_n] \stackrel{\text{def}}{=} Q$.

Guards: process expressions may now be *guarded*, meaning that the behaviour
specified by the process expression is only present if the guard, which may
feature references to parameters, evaluates to true given the current parameter values.

Value-Passing: values may now be communicated between sequential components via activities.

Additional features could have been added, but they would not have greatly
strengthened the usefulness of the language. In the next section, we present
a mapping from PEPA$_k$ to PEPA, and illustrate that quite deliberately, the
underlying algebra has not been changed.

4.1   A Semantics for PEPA$_k$

In this section, we provide a PEPA semantics for PEPA$_k$. We do this so that
the behavior of PEPA$_k$ models within Möbius can be understood formally, and
it is important to note that this translation is not carried out within the tool
itself. Before we give our semantics, we give a preliminary definition that is used
to construct PEPA cooperation and hiding sets.

Definition 2 (Refining data-passing activities) *Let L be a set of action
types, and T be the set of Möbius state variable types (from Section 2.1). The
function* refine(L) *is defined as:*

$$L \cup \{\alpha_i : \alpha \in L, i \in T\}$$

As is conventional, we map all α-activities used in value passing to T-indexed α-
activities. Due to the construction of T, all such refined cooperation and hiding
sets remain countable. Now we can define a PEPA semantics for PEPA$_k$.

Definition 3 (PEPA Semantics for PEPA$_k$) *Let* eval(e, E) *represent the
simple evaluation of expression e in an environment E. Then $\|.\|_E$ is a func-
tion that given an environment E mapping variables to values, maps a PEPA$_k$*

*process or defining equation to a set of PEPA processes or defining equations as
follows:*

$$\|P[\underline{x}] \stackrel{\text{def}}{=} Q\|_E = \{P_{\underline{i}} \stackrel{\text{def}}{=} \|Q\|_{E'} : i_1, \dots, i_n \in T, E' = E[i_1/x_1, \dots, i_n/x_n]\}$$

$$\|P \underset{L}{\bowtie} Q\|_E = \|P\|_E \underset{\text{refine}(L)}{\bowtie} \|Q\|_E$$

$$\|P/L\|_E = \|P\|_E/\text{refine}(L)$$

$$\|P + Q\|_E = \begin{cases} \|P\|_E & \text{if } \|Q\|_E = 0 \\ \|Q\|_E & \text{if } \|P\|_E = 0 \\ \|P\|_E + \|Q\|_E & \text{otherwise} \end{cases}$$

$$\|\text{if } b \text{ then } P\|_E = \begin{cases} \|P\|_E & \text{if eval}(b, E) = \text{true} \\ 0 & \text{otherwise} \end{cases}$$

$$\|(\alpha!e, r).Q\|_E = (\alpha_{\text{eval}(e,E)}, \text{eval}(r, E)).\|Q\|_E$$

$$\|(\alpha?\mathbf{x}, r).Q\|_E = \sum_{\{\alpha_i : i \in T\}} (\alpha_i, \text{eval}(r, E)).\|Q\|_{E[i/x]}$$

$$\|(\alpha, r).Q\|_E = (\alpha, \text{eval}(r, E)).\|Q\|_E$$

$$\|A[\underline{e}]\|_E = A_{\text{eval}(e_1, E), \dots, \text{eval}(e_n, E)}$$

A guarded PEPA_k process P is mapped to P if the guard is true in the current
environment; otherwise it is mapped to the deadlocked process. Deadlocked processes are then removed if they are found to appear in choice contexts. Of course
it is still possible to write a deadlocked PEPA_k process, just as it is possible to
write a deadlocked PEPA process. Parameterized process definitions are mapped
to a set of indexed definitions for each possible combination of parameter values, and process constants provide a way for the PEPA_k process to change the
values of variables itself. A PEPA_k process of the form $(\alpha?\mathbf{x}, r).P$ is mapped to
a sum (choice) over all α_i-guarded copies of P, for $i \in T$ (recall that the set T is
arbitrarily large but finite). A PEPA_k process of the form $(\alpha!e, s).P$ is mapped
to a single PEPA process, specifically $(\alpha_j, s).P$, where j is the result of evaluating e in the environment E. This means that a PEPA_k process of the form
$(\alpha?\mathbf{x}, r).P \underset{\{\alpha\}}{\bowtie} (\alpha!e, s).P$ would be capable of one transition, via a single activity
of type α_j. This has the effect of setting the value of x in the left subcomponent
to be equal to j. This scheme also means that $(\alpha!f, r).P \underset{\{\alpha\}}{\bowtie} (\alpha!e, s).P$ is deadlocked unless $\text{eval}(e, E) = \text{eval}(f, E)$. This is reasonable, with the interpretation
that if both subcomponents are trying to force the setting of a variable in another subcomponent, then they must agree on the value for the variable. This
scheme has been used in previous parameterized process algebras; for example,
see [11,12].

Our chosen semantics is not suitable for understanding a PEPA_k process
such as $P \stackrel{\text{def}}{=} (\alpha?\mathbf{x}, r).P'$ in isolation. As described in [13], the choice over all
α_i-guarded copies of P' means that the sojourn time in state P is not $1/r$
as would be expected, but rather decreases arbitrarily. However, this does not
cause problems for our implementation. We insist that all PEPA_k processes

specified evolve such that if an input activity of type α is enabled, it must be within a cooperation context that enables an output activity of type α. If, at any point during its evolution, the PEPA$_k$ process represented by the model's system equation enables an unmatched input activity, then the model is in error. This should be considered in the same light as the specification of a model with a deadlock; it is perfectly possible to write an incorrect model specification. In our implementation, Möbius catches this error during model execution and halts. It can be shown that for any PEPA$_k$ model satisfying the condition given above, our semantics lead to a PEPA model with an equivalent performance model. Alternative semantic models do exist for value-passing algebras; for example, the STGLA model [13] avoids branching problems by maintaining process variables and expressions in a symbolic form.

Below we give a PEPA$_k$ model of a simple $M/M/s/n$ queue and illustrate the translation to PEPA.

Example 1.

$$Queue[m, s, n] \stackrel{\text{def}}{=} \text{if } (m < n) \text{ then } (\text{in}, \lambda).Queue[m + 1, s, n]$$
$$+ \text{ if } (m > 0) \text{ then } (\text{out}, \mu * \min(s, m)).Queue[m - 1, s, n]$$

translates to a set of definitions over values of s and n, including the following:

$$Queue_{0,s,n} \stackrel{\text{def}}{=} (\text{in}, \lambda).Queue_{1,s,n}$$
$$Queue_{i,s,n} \stackrel{\text{def}}{=} (\text{in}, \lambda).Queue_{i+1,s,n} + (\text{out}, \mu * i).Queue_{i-1,s,n} \text{ for } 0 < i < s$$
$$Queue_{i,s,n} \stackrel{\text{def}}{=} (\text{in}, \lambda).Queue_{i+1,s,n} + (\text{out}, \mu * s).Queue_{i-1,s,n} \text{ for } s \leq i < n$$
$$Queue_{n,s,n} \stackrel{\text{def}}{=} (\text{out}, \mu * m).Queue_{n-1,s,n}$$

We have presented PEPA$_k$, an extension to PEPA, and shown that while expressiveness has been improved and additional modeling flexibility has been provided, the underlying process algebra has not changed. In the next section, we present our application of PEPA$_k$ to the identification of Möbius state variables.

5   Mapping a PEPA$_k$ Process to the AFI

We present a practical technique for identifying state variables in an SPA model, and provide a formal mapping to the Möbius AFI. State variables will be given by PEPA$_k$ process parameters. This means that the modeler will provide explicit guidance on exactly what state may be shared, and means that the modeler will create the PEPA model in such a way that the effect of a change in shared state will be well understood. In order to list the state variables of a PEPA$_k$ process, we first provide the definition of an auxiliary function that calculates the state variables of a sequential PEPA$_k$ process.

Definition 4 (State variables of a sequential PEPA$_k$ process) *The state variables of a sequential PEPA$_k$ process P are given by* svars$(P, \emptyset) \in X$, *where* svars$(., W)$ *is defined as follows:*

$$\text{svars}(a.S, W) = \text{svars}(S, W)$$
$$\text{svars}(S + T, W) = \text{svars}(S, W) \cup \text{svars}(T, W)$$
$$\text{svars}(\text{if } b \text{ then } S, W) = \text{svars}(S, W)$$
$$\text{svars}(A[\underline{e}], W) = \begin{cases} (\{\underline{x}\} \cup \text{svars}(S, W \cup (A, n)) & \text{if } A[\underline{x}] \stackrel{\text{def}}{=} S \text{ and } (A, n) \notin W \\ \emptyset & \text{otherwise} \end{cases}$$

This definition deliberately fails to distinguish between a parameter x of process S and another parameter with the same name of process T, if $T \in ds(S)$. We consider these to represent the same state variable. This mechanism allows the PEPA$_k$ modeler to change the value of a state variable simply; for example, if S and T both have a parameter named x then $T[x] \stackrel{\text{def}}{=} (\alpha, r).S[f(x)]$ will perform an activity (α, r), and immediately afterwards change x's current value from a to $f(a)$. Now we give the state variables for a PEPA$_k$ process, P.

Definition 5 (State variables of a PEPA$_k$ process) *Let $P = \langle S_1, \ldots, S_n \rangle_P$ be a PEPA$_k$ process in vector form. The state variables of P are given by:*

$$\bigcup_{1 \leq i \leq n} \{x_i : x \in \text{svars}(S_i, \emptyset)\}$$

Each state variable is instrumented with an index according to its position in the vector, to ensure that duplicate names do not clash. This completes the definition of the AFI state variables. However, the state variables alone do not characterize the state of the PEPA$_k$ process; it is necessary to take into account the current PEPA$_k$ term too. Just as described in Section 3.1, the process state can be captured using a vector notation. This state is maintained for interacting with model solvers, but is not exported for use in state-sharing or the definition of performance variables. The complete state of a PEPA$_k$ process is discussed further in Section 5.1. In order to complete the mapping to the AFI, we must generate from the PEPA$_k$ process a set of Möbius actions. There are several ways in which this can be done. Given a PEPA$_k$ process $P = \langle S_1, \ldots, S_n \rangle_P$, there are two possibilities for the AFI actions:

- for every $P' \in ds(P)$, the enabled activities of P' (distinguishing duplicates). We reject this idea since it would require that we generate the entire state space of the underlying process in order to provide the mapping.
- for $1 \leq i \leq n$, for every $S_i' \in ds(S_i)$, the enabled activities of S' (distinguishing duplicates). This is reasonable, and can be implemented efficiently. One drawback is in the consideration of impulse rewards; if three sequential components cooperate over an activity, a reward could not be specified for the firing of an activity, as the result of a cooperation between a particular two of the three components.

Instead, we chose a method that can be efficiently computed and does not sacrifice expressibility. Definition 6 specifies the type of Möbius actions.

Definition 6 (Set of Actions) *The set MA is the least set inductively defined as follows:*

- $(a, P) \in MA$ *for all activities* a *and* $PEPA_k$ *processes* P
- *if* $B \in MA$ *then* $(a, P, B) \in MA$
- *if* $B_1, B_2 \in MA$ *then* $(a, P, B_1 \uplus B_2) \in MA$

Now let P be a $PEPA_k$ process. The AFI actions of P are given by $\mathsf{actions}(P, \emptyset) \in MA$, where $\mathsf{actions}(., W)$ is defined as follows:

Definition 7 (Actions for a PEPA$_k$ process) *Let* P *be a* $PEPA_k$ *process. The AFI actions of P are given by* $\mathsf{actions}(P, \emptyset)$, *where* $\mathsf{actions}(., W)$ *is defined as follows:*

$$\mathsf{actions}(S + T, W) = \mathsf{actions}(S, W) \cup \mathsf{actions}(T, W)$$
$$\mathsf{actions}((\alpha?\mathbf{x}, r).S, W) = \{((\alpha, r), S)\} \cup \mathsf{actions}(S, W)$$
$$\mathsf{actions}((\alpha!\mathbf{e}, r).S, W) = \{((\alpha, r), S)\} \cup \mathsf{actions}(S, W)$$
$$\mathsf{actions}((\alpha, r).S, W) = \{((\alpha, r), S)\} \cup \mathsf{actions}(S, W)$$
$$\mathsf{actions}(\text{if } b \text{ then } S, W) = \mathsf{actions}(S, W)$$

$$\mathsf{actions}(A[\underline{e}], W) = \begin{cases} \mathsf{actions}(P, W \cup (A, n)) & \text{if } A[\underline{x}] \stackrel{\text{def}}{=} P \text{ and } (A, n) \notin W \\ \emptyset & \text{otherwise} \end{cases}$$

let $\Phi_P = \mathsf{actions}(P, W)$ *and* $\Phi_Q = \mathsf{actions}(Q, W)$; *then*

$$\mathsf{actions}(P \bowtie_L Q, W) = \{((\alpha, r), P' \bowtie_L Q, B) : B = ((\alpha, r), P', A) \in \Phi_P, \alpha \notin L\} \cup$$
$$\{((\alpha, r), P \bowtie_L Q', B) : B = ((\alpha, r), Q', A) \in \Phi_Q, \alpha \notin L\} \cup$$
$$\{((\alpha, R), P' \bowtie_L Q', B_1 \uplus B_2) : (\alpha, r).P' \bowtie_L (\alpha, s).Q' \xrightarrow{(\alpha, R)},$$
$$B_1 = ((\alpha, r), P', A_1) \in \Phi_P,$$
$$B_2 = ((\alpha, s), Q', A_2) \in \Phi_Q\}$$
$$\mathsf{actions}(P/L, W) = \{((\tau, r), P', B) : B = ((\alpha, r), P', A) \in \Phi_P, \alpha \in L\} \cup$$
$$\{((\alpha, r), P', B) : B = ((\alpha, r), P, A) \in \Phi_P, \alpha \notin L\}$$

For each derivative of each sequential component, this function computes every enabled activity. The AFI actions of a cooperation $P \bowtie_L Q$ are

- actions associated with individual behavior of each subcomponent (types that do not match those in L).
- for each pair of AFI actions with types $\alpha \in L$, a new action consisting of a cooperation between the two.

Thus a Möbius action a is a structure consisting of a PEPA activity, the derivative that results from its completion, and then a set of the Möbius actions of the model's subcomponents that have combined to form a. The advantage of this technique is that every activity that could be enabled by any derivative of P is mapped to a Möbius AFI action. This means that the Möbius modeler can distinguish and assign an impulse reward to a composite activity resulting from the evolution of a chosen subset of model subcomponents. The disadvantage of this technique is that since we directly compute "products" of activities, there may be a proliferation of AFI actions that correspond to PEPA activities that may never be enabled.

We have presented a mapping from $PEPA_k$ to the Möbius AFI. This means that $PEPA_k$ models can be composed with other Möbius atomic models using equivalence sharing. The ability of another model to unilaterally change some shared state has some consequences for the parameterized process, which we discuss next.

5.1   Implications of Modeling with $PEPA_k$

We have described a mapping from $PEPA_k$ to the AFI, and have shown that $PEPA_k$ is no more powerful than PEPA itself. By providing the modeler with some of the convenience of a programming language, we

- facilitate the construction of concise $PEPA_k$ specifications that have a natural PEPA semantics.
- cause the modeler to structure his definitions in such a way that a change in the value of a state variable ($PEPA_k$ process parameter) due to a partner model will cause a meaningful and understandable change in the state of the $PEPA_k$ model.

The last point is an important one, and justifies our selection of this method for implementation. The addition of such "cues" into the PEPA model makes equivalence sharing meaningful and useful.

In Section 5, we stated that the PEPA formalism maps process parameters to state variables for use in state sharing and the definition of performance variables, but that the state of the $PEPA_k$ process is also maintained and communicated to model solvers via the AFI. This means that we can construct a $PEPA_k$ model P that Möbius interprets as having a larger state space than $\|P\|_\emptyset$. Consider the following $PEPA_k$ definition:

$$S[x] \stackrel{def}{=} \text{ if } x \neq 1 \text{ then } (\alpha_1, r).(\beta, s).S[1]$$
$$+ \text{ if } x \neq 2 \text{ then } (\alpha_2, r).(\beta, s).S[2]$$
$$+ \text{ if } x \neq 3 \text{ then } (\alpha_3, r).(\beta, s).S[3]$$

Translating this to PEPA leads to:

$$S_1 \stackrel{def}{=} (\alpha_2, r).(\beta, s).S_2 + (\alpha_3, r).(\beta, s).S_3$$
$$S_2 \stackrel{def}{=} (\alpha_1, r).(\beta, s).S_1 + (\alpha_3, r).(\beta, s).S_3$$
$$S_3 \stackrel{def}{=} (\alpha_1, r).(\beta, s).S_1 + (\alpha_2, r).(\beta, s).S_2$$

The PEPA process has 6 states. However, using the AFI, the Möbius state space generator will detect 9 unique states for the $PEPA_k$ process. The reason for this is that the process itself can be in the state $(\beta, s).S[1]$ while the state variable x may either have the value 2 or the value 3 (and similarly for the other derivatives of $S[x]$). These states are equal by every reasonable process algebra equivalence, since $(\beta, s).S[1]$ makes no further use of the value of x. However, what is crucially different is that the AFI now allows a partner model to use the value of x while the $PEPA_k$ model is in this state. If the modeler wishes to generate the smallest reasonable state space, the model could alternatively be specified as below:

$$S[x] \stackrel{\text{def}}{=} \text{if } x \neq 1 \text{ then } (\alpha_1, r).S'[1]$$
$$+ \text{ if } x \neq 2 \text{ then } (\alpha_2, r).S'[2]$$
$$+ \text{ if } x \neq 3 \text{ then } (\alpha_3, r).S'[3]$$
$$S'[x] \stackrel{\text{def}}{=} (\beta, s).S[x]$$

This works because the value of the state variable is changed one activity sooner. However, this process will not behave identically to its original partner process. Furthermore, although in isolation, the $PEPA_k$ model has a state space consistent with the translated PEPA model. In an equivalence sharing relationship, the partner model still has the opportunity to change the value of x at any point, leading in this case to a slightly larger state space. If the modeler aims to employ his $PEPA_k$ model in an equivalence sharing relationship, this is a novel issue of which he must be aware.

One interesting aspect of this work is the extent to which equivalence relations may be employed for aggregation when equivalence sharing is being used. We can certainly ensure that $P \bowtie_L P'$ is treated as equivalent to $P' \bowtie_L P$ and check for this as detailed in [14]. However, if P is defined with a parameter, e.g., $P[x] \stackrel{\text{def}}{=} (\alpha, r).P'[x + 1]$, then $P[a] \bowtie_L P'[b]$ will export two state variables, x_1 and x_2, with values a and b respectively, via the AFI. A partner model may be relying on the individual values of these variables, and thus it would be incorrect to equate this process to $P'[b] \bowtie_L P[a]$. For processes with parameters, the order in which they appear in the static structure of a term must be preserved.

6   Example

We present an example to illustrate the implementation and use of $PEPA_k$ in Möbius. Our model is of a simple factory system that consists of three robot arms. Two robots are responsible for removing items from a conveyor belt, packaging several items together into two units and then depositing one assembled unit into each of the two bins. The third robot removes assembled units from the shared bins and places them on another conveyor belt for later processing. It does this by always attempting to remove an item from the first bin if it can, and only if it cannot, removing and passing along an item from the second bin.

The first two robot arms behave identically, and are modeled by one SAN as shown in Figure 2a. The leftmost activity fires until place Accumulate is full;

meanwhile, the robot arm assembles units, incrementing the values of places Bin1 and Bin2 when activity Assemble fires. The PEPA$_k$ model of the third robot arm is shown in Figure 2b. From the specification of process $Consume[b1, b2]$, it can be seen that activity (outb1, $b1r$) is enabled if the value of parameter $b1$ is positive, and (outb2, $b2r$) is enabled if $b2$ is positive, and $b1$ equals zero. From here, the

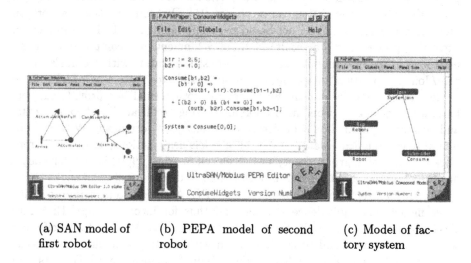

(a) SAN model of first robot

(b) PEPA model of second robot

(c) Model of factory system

Fig. 2. Example models

overall model of the system can be easily built, and is shown in Figure 2c. We use the Replicate/Join formalism; two copies of the first robot are produced by applying a replicate node to the SAN model. The Replicate creates an integer number of copies of a model, and does not distinguish individual copies. We insist that both copies of the first robot share places Bin1 and Bin2. Next we use a Join node to apply equivalence sharing to the Replicated model of the first robot, and the PEPA$_k$ model of the second robot. The Möbius tool allows us to specify that shared place Bin1 should be further shared (identified) with the parameter $b1$ of the PEPA model, and similarly that Bin2 should be identified with $b2$. In this way, we create an accurate model of the system as a whole. Due to our mapping to the AFI, it is now possible to use the Möbius tool's analytical solvers or discrete-event simulator to investigate the behavior of this model over time.

7 Future Work and Conclusion

One area of future work is in the design of an *action-sharing* composition modeling formalism for Möbius. Here, Möbius can benefit from the experience and results of the SPA community. PEPA, and SPAs in general, are packaged with a compositional theory based around synchronization via actions. In contrast, the Möbius framework, and its implementation as the Möbius tool, currently supports the synchronization of concurrent models via shared state. One reasonable and useful extension to the theory would be to allow action-synchronization operators, such as PEPA's cooperation, to be applicable to any submodels that satisfy the AFI. Many choices of operator may be useful; for example, the operator may insist that some particular subset of "cooperating" activities complete before the submodels change state. The literature features several ways in which the rate of cooperation can be chosen [15]. Furthermore, there is work on building compositional Petri net-based modeling languages [16,17], and also on exploiting process algebra results in the development of a compositional Petri net theory [18]. By combining such action-based operators with Möbius's Replicate/Join composition formalism, it will be possible to create a new and general model composition formalism. Furthermore, both formalisms provide support for state space aggregation based upon identifying "replicated" subcomponents, and this aggregation should be preserved in any joint formalism. With the potential to communicate data between submodels using actions, and with current work on exploiting group theory to detect symmetries [19], this new formalism has the potential to be fruitful in both theory and practice.

We have presented a method of allowing PEPA models to be composed with other models via equivalence sharing, and thus for incorporating PEPA into the Möbius tool. As a result, the Möbius tool may now be used to specify and solve PEPA models, and to combine them with existing modeling formalisms, such as SANs and an extended Markov chain formalism called Buckets and Balls [20]. We believe this illustrates the flexibility and generality of both the Möbius framework and the AFI. Furthermore, we expect that a similar mapping can be developed for SPAs other than PEPA.

Acknowledgments. We would like to thank Jane Hillston for her valuable suggestions and for making this collaboration possible. We also thank Jay Doyle and Dan Deavours for their patient explanations and assistance, and Holger Hermanns for several useful discussions. Thanks also to the other members of the Möbius group: Tod Courtney, David Daly, Salem Derisavi, and Patrick Webster.

References

1. J. M. Doyle, "Abstract model specification using the Möbius modeling tool," M.S. thesis, University of Illinois at Urbana-Champaign, 2000.
2. G. Clark, T. Courtney, D. Daly, D. Deavours, S. Derisavi, J. M. Doyle, W. H. Sanders, and P. Webster, "The Möbius modeling tool," in *Proc. of PNPM'01: 10th International Workshop on Petri Nets and Performance Models, Aachen, Germany (to appear)*, September 2001.
3. J. F. Meyer, A. Movaghar, and W. H. Sanders, "Stochastic activity networks: Structure, behavior and applications," *Proc. International Workshop on Timed Petri Nets*, pp. 106–115, 1985.
4. J. Hillston, *A Compositional Approach to Performance Modelling*, Cambridge University Press, 1996.

5. W. H. Sanders, "Integrated frameworks for multi-level and multi-formalism modeling," in *Proc. PNPM'99: 8th International Workshop on Petri Nets and Performance Models, Zaragoza, Spain*, September 1999, pp. 2–9.

6. D. Deavours and W. H. Sanders, "Möbius: Framework and atomic models," in *Proc. PNPM'01: 10th International Workshop on Petri Nets and Performance Models, Aachen, Germany (to appear)*, September 2001.

7. W. H. Sanders, W. D. Obal II, M. A. Qureshi, and F. K. Widjanarko, "The *ultrasan* modeling environment," *Performance Evaluation*, vol. 24, no. 1, pp. 89–115, October-November 1995.

8. W. H. Sanders and J. F. Meyer, "Reduced base model construction methods for stochastic activity networks," *IEEE Journal on Selected Areas in Communications*, vol. 9, no. 1, pp. 25–36, Jan. 1991.

9. D. Montgomery, *Design and Analysis of Experiments*, John Wiley & Sons, Inc., 5th edition, 2001.

10. R. Milner, *Communication and Concurrency*, International Series in Computer Science. Prentice Hall, 2nd edition, 1989.

11. T. Bolognesi and E. Brinksma, "Introduction to the ISO specification language LOTOS," *Computer Networks and ISDN Systems*, vol. 14, pp. 25–59, 1987.

12. H. Hermanns and M. Rettelbach, "Toward a superset of basic LOTOS for performance prediction," *Proc. of 4th Workshop on Process Algebras for Performance Modelling (PAPM)*, pp. 77–94, 1996.

13. M. Bernardo, *Theory and Application of Extended Markovian Process Algebra*, Ph.D. thesis, University of Bologna, Italy, 1999.

14. S. Gilmore, J. Hillston, and M. Ribaudo, "An efficient algorithm for aggregating PEPA models," *IEEE Transactions on Software Engineering*, 2001.

15. J. Hillston, "The nature of synchronisation," *Proc. of 2nd Workshop on Process Algebras for Performance Modelling (PAPM)*, pp. 51–70, 1994.

16. S. Donatelli, "Superposed generalized stochastic Petri nets: Definition and efficient solution," in *Application and Theory of Petri Nets 1994, Lecture Notes in Computer Science 815 (Proc. 15th International Conference on Application and Theory of Petri Nets, Zaragoza, Spain)*, R. Valette, Ed., pp. 258–277. Springer-Verlag, June 1994.

17. I. Rojas, *Compositional Construction of SWN Models*, Ph.D. thesis, The University of Edinburgh, 1997.

18. E. Best, R. Devillers, and M. Koutny, *Petri Net Algebra*, Monographs in Theoretical Computer Science. An EATCS Series. Springer-Verlag, 2000.

19. W. D. Obal II, *Measure-Adaptive State-Space Construction Methods*, Ph.D. thesis, The University of Arizona, 1998.

20. W. J. Stewart, *Introduction to the Numerical Solution of Markov Chains*, Princeton University Press, 1994.

Author Index

Lecture Notes in Computer Science

For information about Vols. 1–2084
please contact your bookseller or Springer-Verlag

Vol. 2124: W. Skarbek (Ed.), Computer Analysis of Images and Patterns. Proceedings, 2001. XV, 743 pages. 2001.

Vol. 2125: F. Dehne, J.-R. Sack, R. Tamassia (Eds.), Algorithms and Data Structures. Proceedings, 2001. XII, 484 pages. 2001.

Vol. 2126: P. Cousot (Ed.), Static Analysis. Proceedings, 2001. XI, 439 pages. 2001.

Vol. 2127: V. Malyshkin (Ed.), Parallel Computing Technologies. Proceedings, 2001. XII, 516 pages. 2001.

Vol. 2129: M. Goemans, K. Jansen, J.D.P. Rolim, L. Trevisan (Eds.), Approximation, Randomization, and Combinatorial Optimization. Proceedings, 2001. IX, 297 pages. 2001.

Vol. 2130: G. Dorffner, H. Bischof, K. Hornik (Eds.), Artificial Neural Networks – ICANN 2001. Proceedings, 2001. XXII, 1259 pages. 2001.

Vol. 2132: S.-T. Yuan, M. Yokoo (Eds.), Intelligent Agents. Specification. Modeling, and Application. Proceedings, 2001. X, 237 pages. 2001. (Subseries LNAI).

Vol. 2136: J. Sgall, A. Pultr, P. Kolman (Eds.), Mathematical Foundations of Computer Science 2001. Proceedings, 2001. XII, 716 pages. 2001.

Vol. 2138: R. Freivalds (Ed.), Fundamentals of Computation Theory. Proceedings, 2001. XIII, 542 pages. 2001.

Vol. 2139: J. Kilian (Ed.), Advances in Cryptology – CRYPTO 2001. Proceedings, 2001. XI, 599 pages. 2001.

Vol. 2141: G.S. Brodal, D. Frigioni, A. Marchetti-Spaccamela (Eds.), Algorithm Engineering. Proceedings, 2001. X, 199 pages. 2001.

Vol. 2142: L. Fribourg (Ed.), Computer Science Logic. Proceedings, 2001. XII, 615 pages. 2001.

Vol. 2143: S. Benferhat, P. Besnard (Eds.), Symbolic and Quantitative Approaches to Reasoning with Uncertainty. Proceedings, 2001. XIV, 818 pages. 2001. (Subseries LNAI).

Vol. 2146: J.H. Silverman (Eds.), Cryptography and Lattices. Proceedings, 2001. VII, 219 pages. 2001.

Vol. 2147: G. Brebner, R. Woods (Eds.), Field-Programmable Logic and Applications. Proceedings, 2001. XV, 665 pages. 2001.

Vol. 2149: O. Gascuel, B.M.E. Moret (Eds.), Algorithms in Bioinformatics. Proceedings, 2001. X, 307 pages. 2001.

Vol. 2150: R. Sakellariou, J. Keane, J. Gurd, L. Freeman (Eds.), Euro-Par 2001 Parallel Processing. Proceedings, 2001. XXX, 943 pages. 2001.

Vol. 2151: A. Caplinskas, J. Eder (Eds.), Advances in Databases and Information Systems. Proceedings, 2001. XIII, 381 pages. 2001.

Vol. 2152: R.J. Boulton, P.B. Jackson (Eds.), Theorem Proving in Higher Order Logics. Proceedings, 2001. X, 395 pages. 2001.

Vol. 2153: A.L. Buchsbaum, J. Snoeyink (Eds.), Algorithm Engineering and Experimentation. Proceedings, 2001. VIII, 231 pages. 2001.

Vol. 2154: K.G. Larsen, M. Nielsen (Eds.), CONCUR 2001 – Concurrency Theory. Proceedings, 2001. XI, 583 pages. 2001.

Vol. 2157: C. Rouveirol, M. Sebag (Eds.), Inductive Logic Programming. Proceedings, 2001. X, 261 pages. 2001. (Subseries LNAI).

Vol. 2158: D. Shepherd, J. Finney, L. Mathy, N. Race (Eds.), Interactive Distributed Multimedia Systems. Proceedings, 2001. XIII, 258 pages. 2001.

Vol. 2159: J. Kelemen, P. Sosík (Eds.), Advances in Artificial Life. Proceedings, 2001. XIX, 724 pages. 2001. (Subseries LNAI).

Vol. 2161: F. Meyer auf der Heide (Ed.), Algorithms – ESA 2001. Proceedings, 2001. XII, 538 pages. 2001.

Vol. 2162: Ç. K. Koç, D. Naccache, C. Paar (Eds.), Cryptographic Hardware and Embedded Systems – CHES 2001. Proceedings, 2001. XIV, 411 pages. 2001.

Vol. 2164: S. Pierre, R. Glitho (Eds.), Mobile Agents for Telecommunication Applications. Proceedings, 2001. XI, 292 pages. 2001.

Vol. 2165: L. de Alfaro, S. Gilmore (Eds.), Process Algebra and Probabilistic Methods. Proceedings, 2001. XII, 217 pages. 2001.

Vol. 2166: V. Matoušek, P. Mautner, R. Mouček, K. Taušer (Eds.), Text, Speech and Dialogue. Proceedings, 2001. XIII, 452 pages. 2001. (Subseries LNAI).

Vol. 2170: S. Palazzo (Ed.), Evolutionary Trends of the Internet. Proceedings, 2001. XIII, 722 pages. 2001.

Vol. 2172: C. Batini, F. Giunchiglia, P. Giorgini, M. Mecella (Eds.), Cooperative Information Systems. Proceedings, 2001. XI, 450 pages. 2001.

Vol. 2176: K.-D. Althoff, R.L. Feldmann, W. Müller (Eds.), Advances in Learning Software Organizations. Proceedings, 2001. XI, 241 pages. 2001.

Vol. 2177: G. Butler, S. Jarzabek (Eds.), Generative and Component-Based Software Engineering. Proceedings, 2001. X, 203 pages. 2001.

Vol. 2181: C. Y. Westort (Eds.), Digital Earth Moving. Proceedings, 2001. XII, 117 pages. 2001.

Vol. 2184: M. Tucci (Ed.), Multimedia Databases and Image Communication. Proceedings, 2001. X, 225 pages. 2001.

Vol. 2186: J. Bosch (Ed.), Generative and Component-Based Software Engineering. Proceedings, 2001. VIII, 177 pages. 2001.

Vol. 2188: F. Bomarius, S. Komi-Sirviö (Eds.), Product Focused Software Process Improvement. Proceedings, 2001. XI, 382 pages. 2001.

Vol. 2189: F. Hoffmann, D.J. Hand, N. Adams, D. Fisher, G. Guimaraes (Eds.), Advances in Intelligent Data Analysis. Proceedings, 2001. XII, 384 pages. 2001.

Vol. 2190: A. de Antonio, R. Aylett, D. Ballin (Eds.), Intelligent Virtual Agents. Proceedings, 2001. VIII, 245 pages. 2001. (Subseries LNAI).

Vol. 2191: B. Radig, S. Florczyk (Eds.), Pattern Recognition. Proceedings, 2001. XVI, 452 pages. 2001.

Vol. 2193: F. Casati, D. Georgakopoulos, M.-C. Shan (Eds.), Technologies for E-Services. Proceedings, 2001. X, 213 pages. 2001.